# English for IT Communication

*English for IT Communication* provides a comprehensive introduction for students and professionals studying IT or computer science and covers all forms of technical communication from emails and memos through procedures to reports and design specs. In each case, the book offers multiple real-world examples, looking at who the texts are written for, what their purpose is, and how these affect what is on the page.

Key features of this book include

- How to write for different audiences and purposes
- How to design documents for ease of access and understanding
- How to communicate in multimodal media
- How to reference in IEEE
- Multiple different examples and breakdowns of common text types to show how they are written and to produce an understanding of quality in each
- Online support material including authentic examples of different workplace genres and a reference section covering relevant research studies and weblinks for readers to better understand the topics covered in each chapter
- Internationalized coverage of IT communication exemplars

This book is an accessible guide to writing effective forms of IT communications of the kind needed for all IT degree programmes which aim to prepare students for the modern workplace. Practical and clearly written, it is designed to introduce readers to features of the most common genres in IT and computer science.

**Tony Myers** has worked for over 15 years in the tertiary sector helping learners develop skills in academic and technical writing. He is currently an assistant professor teaching at Zayed University, specializing in ESAP. He has previously written books on academic writing, psychoanalysis, and literature for Routledge and for Bloomsbury Academic and has published articles on genre pedagogy, feedback literacy, and remote learning.

**Jaime Buchanan** is a senior lecturer at Zayed University who has taught academic and technical writing in higher education since 2010. She has published articles on subjects as diverse as pedagogy, feedback literacy, and remote learning. She is passionate about developing an awareness of effective writing strategies in her students to promote best practice communication in IT.

## Routledge Applied English Language Introductions
Series Editor: Prithvi Shrestha

*The Open University, UK*

**Routledge Applied English Language Introductions** is a series of practical and accessible textbooks which introduce spoken and written English for a specific discipline, covering elements such as academic writing, email communication, presentations, writing up reports, and job applications.

The books are designed for students and professionals who aim to refine their English skills for their chosen field and professional career.

### English for Business Communication
*Mable Chan*

### English for IT Communication
*Tony Myers and Jaime Buchanan*

More information about this series can be found at www.routledge.com/Routledge-Applied-English-Language-Introductions/book-series/RAELIS.

# English for IT Communication

Tony Myers and Jaime Buchanan

Routledge
Taylor & Francis Group

LONDON AND NEW YORK

Designed cover image: Getty

First published 2025
by Routledge
4 Park Square, Milton Park, Abingdon, Oxon OX14 4RN

and by Routledge
605 Third Avenue, New York, NY 10158

*Routledge is an imprint of the Taylor & Francis Group, an informa business*

© 2025 Tony Myers and Jaime Buchanan

The right of Tony Myers and Jaime Buchanan to be identified as authors of this work has been asserted in accordance with sections 77 and 78 of the Copyright, Designs and Patents Act 1988.

*British Library Cataloguing-in-Publication Data*
A catalogue record for this book is available from the British Library

ISBN: 978-1-032-64750-0 (hbk)
ISBN: 978-1-032-64749-4 (pbk)
ISBN: 978-1-032-64752-4 (ebk)

DOI: 10.4324/9781032647524

Typeset in Times New Roman
by Apex CoVantage, LLC

Support Material for Instructors is available at www.routledge.com/9781032647494

For L and G – the original audience

# Contents

*Acknowledgements*                                                          *xiv*

**1    Audience and purpose in IT**                                              1
    *1.1 Introduction to audience and purpose  1*
        *What is IT communication, anyway?  1*
        *Attributes of effective technical communication  3*
    *1.2 Analyzing audience: who's going to read this?  5*
        *Imbalances  6*
    *1.3 Common types of IT audience  7*
    *1.4 Analyzing purpose: why are they reading it?  7*
        *Persuasion in technical communication  9*
    *1.5 Language for audience and purpose  13*
        *Formality  13*
    *1.6 Review  14*
        *Reflection questions  15*
        *Application tasks  15*
    *Works cited  20*

**2    Design principles in IT**                                                21
    *2.1 Introduction to design principles  21*
        *Why are design principles important in technical communication?  21*
        *Skimmable takeaways and easy access points  21*
    *2.2 Design principles  22*
        *Unity  22*
        *Balance  23*
        *Alignment  24*
        *Hierarchy  27*
        *Emphasis  28*
        *Proportion  31*
        *White space  31*
        *Repetition  33*

     *Movement  34*

     *Contrast  36*

  2.3  *Design in IT contexts: headings  37*

     *Content-oriented headings  38*

  2.4  *Design in IT contexts: lists  38*

     *Types of lists  40*

  2.5  *Data visualization strategies  40*

     *Entity relationship diagrams (ERDs)  41*

     *Use case diagrams  42*

     *User flow diagrams  42*

  2.6  *Review  44*

     *Reflection questions  44*

     *Application tasks  44*

  *Works cited  46*

**3    Workplace communication in IT**          47

  3.1  *Introduction to workplace communication  47*

     *Internal vs. external communications  47*

     *Intercultural considerations  49*

  3.2  *Email  51*

     *Audience  51*

     *Purpose  51*

     *Layout  52*

  3.3  *Memos  56*

     *Audience  56*

     *Purpose  57*

     *Layout  57*

     *Recommendation memos  59*

  3.4  *Enterprise social media  61*

     *Audience  61*

     *Purpose  62*

     *Layout  62*

  3.5  *Language focus in workplace communications  63*

     *Using* You *language  63*

     *Not using* You *language  64*

     *Avoid negative phrasing  64*

  3.6  *Review  65*

     *Reflection questions  65*

     *Application tasks  65*

  *Works cited  68*

**4    Process documents in SDLC**          69

  4.1  *Introduction to project management documentation  69*

     *Types of project management  69*

The value of SDLC documents  69
Different types of SDLC document  70
4.2  Work breakdown structures  71
What is a work breakdown structure?  71
Who uses a WBS?  72
Gantt charts  72
Kanban boards  74
4.3  Roadmaps  75
Strategic roadmaps  76
Technology roadmaps  77
Release roadmaps  78
4.4  Other process documents  79
Coding standards  79
Working papers  80
4.5  SDLC documentation best practice  80
SDLC document writing guidelines  80
SDLC document management  81
Docs as code  82
4.6  Review  82
Reflection questions  82
Application tasks  83
Works cited  84

**5    System documents in SDLC**                                                 85
5.1  Introduction to product documentation  85
Two types of product documentation  85
5.2  Product requirement document  85
Audience  85
Purpose  86
Layout  86
User stories  88
5.3  Software requirements specification document  92
Audience  92
Purpose  92
Layout  92
5.4  UX design documentation  94
Audience  94
Purpose  94
Layout  94
5.5  API documentation  95
Audience  96
Purpose  96
Layout  96

5.6  *Quality assurance documentation  99*
    *Audience  99*
    *Purpose  100*
    *Layout  100*
5.7  *Language focus in system documentation  103*
    *Key grammar: grammatical clarity I  103*
    *Key grammar: grammatical clarity II  103*
    *Key grammar: tenses  103*
5.8  *Review  104*
    *Reflection questions  104*
    *Application tasks  104*
*Works cited  107*

**6  User documents in SDLC**    108
6.1  *Introduction to user documentation  108*
6.2  *End-user documentation  108*
    *Importance of writing effective instructions  108*
    *Type of end-user documentation  109*
    *Writing for audience and purpose  109*
    *Instruction manuals and quick-start guides  111*
    *Policy documents and standard operating procedures  117*
    *FAQs and self-access IT resources  121*
    *Some important guidelines for procedural writing  122*
6.3  *Online tools  123*
6.4  *Language focus for user documents  124*
    *Command/imperative voice  124*
    *Sequencing language  124*
    *Discussing the results of an action: two grammatical choices  124*
6.5  *System admin documentation  125*
    *Common types of system admin documentation  125*
6.6  *Review  131*
    *Reflection questions  131*
    *Application tasks  132*
*Works cited  133*

**7  Report writing in IT**    134
7.1  *Introduction to report writing in IT  134*
    *The value of reports  134*
    *Different kinds of reports  134*
7.2  *Proposals  136*
    *Audience  136*
    *Purpose  136*
    *Layout  137*

7.3  Recommendation reports  137
    Audience  138
    Purpose  138
    Layout  138
7.4  Feasibility reports  138
    Audience  139
    Purpose  139
    Layout  139
7.5  Progress reports  140
    Audience  140
    Purpose  140
    Layout  140
7.6  Evaluation reports  141
    Audience  141
    Purpose  141
    Layout  141
7.7  Common report features  142
    Introductions  142
    Criteria and parameters  143
    Money management  145
    Rhetorical support  145
    Executive summaries  147
7.8  Language focus in report writing  148
    Maintaining the information flow  148
    Conveying stance  149
7.9  Review  150
    Reflection questions  150
    Application tasks  150
Works cited  153

8    IEEE referencing and formatting                                      154
8.1  Introduction to IEEE  154
    What is IEEE?  154
    Who uses IEEE?  154
    What is the value of using IEEE?  154
8.2  The basics of referencing  155
    What is a referencing style?  155
    How many referencing styles are there?  155
8.3  In-text citations  156
    How do I cite?  156
    Where do I cite?  156
    What are the different styles of citation?  157
    Why do I use the different styles?  157

*When do I cite? 158*

*When do I not cite? 158*

8.4 *End-reference list 160*

*How do I create a reference list? 160*

*What does each entry need in a reference list? 160*

*Authors 161*

*Titles 161*

*Dates 162*

*Sources 162*

*Referencing internet sources 162*

*Referencing reports 163*

*Referencing standards 163*

*Referencing articles 163*

*Referencing books 164*

8.5 *Abbreviations and locators 164*

*List of common abbreviations 164*

*List of common locators 165*

8.6 *IEEE style guide 166*

*Tables 166*

*Figures 167*

*Headings 168*

8.7 *Drafting tips for organizing source material 168*

*Drafting by hand 169*

*Drafting with software 169*

8.8 *Review 169*

*Reflection questions 169*

*Application tasks 170*

*Works cited 171*

**9 Multimodal communication in IT** 172

9.1 *Introduction to multimodal communication 172*

*What is multimodal communication? 172*

9.2 *Video conferencing 174*

9.3 *Presentations 174*

*The audience for presentations 174*

*The purpose of presentations 176*

*Structuring presentations 176*

*Slide design 178*

*Delivery 183*

*Checklist for presentations 184*

9.4 *Language focus in presentations 184*

*SCQR signposts 184*

*Transition signposts 185*

*Closing language 185*

9.5 *Review  185*
    *Reflection questions  185*
    *Application tasks  185*
*Works cited  189*

**10   Case studies: Applied communication in IT**                                190
*Note to the teacher/independent learner  190*
*Case study 1: conduct a stakeholder profile  190*
    *1   Audience analysis  191*
    *2   Audience variety  191*
    *3   Purpose analysis  192*
    *4   Task analysis  192*
    *Suggested assessment tasks  194*
*Case study 2: redesigning a company website  194*
    *1   Website design  195*
    *2   Modifying website design  195*
*Case study 3: reviewing automated messages  195*
    *1   Automated message analysis  196*
    *Suggested assessment tasks  196*
*Case study 4a: developing a micro-payments app  197*
    *1   Project management model comparative analysis  197*
    *2   Time, scheduling, and cost management documentation  198*
    *3   Feasibility study  198*
    *Suggested assessment tasks  199*
*Case Study 4b: developing a micro-payments app  199*
    *1   Use case diagram  199*
    *2   Software requirements specification document  200*
    *Suggested assessment tasks  201*
*Case study 5: writing for end users  202*
    *1   Extended writing task  202*
    *2   Optional extended writing task  202*
    *Further case studies  202*
*Works cited  203*

*Index*                                                                            *204*

# Acknowledgements

We would like to thank Prithvi Shrestha for his visionary enthusiasm and support, Nick Moore and Mousa Al-kfairy for their insights and suggestions, and Eleni Steck and Bex Hume for their ideas, wise counsel and professionalism.

# Chapter 1

# Audience and purpose in IT

## 1.1 Introduction to audience and purpose

Two of the most important concepts in English for IT students and professionals are audience and purpose. Who you are communicating to and the reason you are communicating inform all of the choices you make. They are two of the foundational concepts underpinning this book, which is why we begin by examining them. Carefully considering your audience and purpose will form the basis of developing effective technical writing skills throughout this book. We use them to explore the interactions of key stakeholders in the software design life cycle (SDLC), which is explored beginning in Chapter 4.

### What is IT communication, anyway?

The Society for Technical Communication defines technical communication in the following way:

> Technical communication involves the delivery of clear, consistent, and factual information – often stemming from complex concepts – for safe and efficient use and effective comprehension by users. Technical communication is a user-centred approach for providing the right information, in the right way, at the right time so that the user's life is more productive. The value that technical communicators deliver is twofold: They make information more usable and accessible to those who need that information, and they advance the goals of the companies and organizations that employ them [1, para 1].

Here, the emphasis is on the usability of technical content by end users. In this book, we will use the term IT communication to focus on how people working in the IT industry communicate with each other.

Traditional models of communication typically portray a sender and a receiver, a choice of channel or mode of communication, and the message itself with the possibility of obstruction of the message [2, 3].

Other models conceive of communication as more interactional in nature. In this sense, it is used to build and maintain relationships. Communication is used to establish rapport and build social connections [4, 5].

A good example of communication that is not informational can be found in how people tend to greet one another in different cultural or social settings. In Arabic, for example, it is common to greet each other with a standard phrase of *Assalaamwa'aleykum* (Peace be upon you).

DOI: 10.4324/9781032647524-1

In some European cultures, it is common to greet people you know with a kiss on one or both cheeks. These are not information exchanges, but they fulfil an important social function. These exchanges serve as social bonding. They are cultural patterns and can be different based on gender, race/ethnicity, class, etc. [6]. In other cultures and contexts, saying hello to people may be optional. It may be perfectly acceptable to pass in a corridor without exchanging any verbal or non-verbal communication.

Being able to identify the needs of your audience will help you to tailor your communication to make it more effective. In many IT contexts, especially in software development, our audience does not really read what we create. Our audience is more likely to consume content (as in absorbing it, much like watching YouTube videos on auto play). Think about how you interact with an app on your device. You engage or interact with content. In this chapter, we will refer to how an audience digests IT-related content using the following verbs:

- Consume
- Engage
- Interact
- Read

It is a good idea to have a solid understanding of each of these words, how they are similar, and the differences in how they are used. If your first language is not English, you might consider translating each into your first language and comparing them there to deepen your understanding of how they are used.

Let's begin by defining these important keywords.

*Purpose* refers to the reason for your communication. What do you hope to achieve by communicating? Are you trying to inform, persuade, or instruct your audience? Are you initiating contact about a topic, or are you responding to a request? Purpose can also serve different functions, like mentioned earlier, in interactional perspectives, purpose can be to strengthen a connection or to acknowledge a hierarchical or power structure within a system or institution.

Different purposes call for different forms (or modes) of communication. For example, an urgent internal workplace message is often best conveyed by telephone or text message, unless having a written record of the exchange is important. More serious workplace messages are communicated through more formal channels, such as policy documents and letters of resignation. In the case of a policy or a resignation, it is important for all people involved to be able to refer to a permanent (written) record.

Another consideration is the language choices you make for the type of message, so when you are clear about your purpose, you can start to think about the best way to achieve it. More discussion of some guidelines for language choices will be discussed later in this chapter and throughout the book.

*Audience* refers to the people you are communicating with. You might be referring to the reader of a design spec (a software engineer), or the listener or viewer of a tutorial explaining how to set up an account. Listening and watching instructions is also very different from reading instructions. Unless you can pause and rewind, you as the reader are not in control of the flow of information (discussed in more detail in Chapter 9). Whether they are reading, watching, or listening, we use the term 'audience' to refer to the intended recipient of your IT message.

An audience can be direct – the person or group to whom the message is directed. A **direct audience** in technical communication typically needs to do something as a result of the message.

For instance, they might need to test the data modelling of an app or approve a proposal for a new data handling policy. **Indirect audiences**, by contrast, do not necessarily need to act as a result of a message. Sometimes they just need to be aware of a situation. Indirect audiences could be supervisors in the workplace who are copied into internal communications, or they could be competitors viewing how your technical writing team handle and respond to user comments and rating on application provider platforms like Google Play.

In technical communication, part of considering your audience means considering their relationship to the message and the situation in which the communication occurs. What are their needs and interests? What do they already know about the topic? What is their level of expertise in this situation or context? By understanding your audience, you can tailor your communication to be more relevant and engaging.

### Attributes of effective technical communication

Technical communication is pragmatic [5]. Unlike non-pragmatic forms of communication such as text messages among friends, pragmatic communication involves a careful consideration of the context within which the communication occurs. Non-pragmatic communication relies on already established context, so it doesn't need to provide much because there is already a considerable amount of shared knowledge and understanding.

**Pragmatic communication either does not already have an established knowledge base or it cannot assume one.** It considers (inter)cultural norms and the relationship between the communicators. Pragmatic communication also involves being aware of your audience's perspective – whether they are likely to have background knowledge, for example, or whether they are new to the situation or task at hand. This also relates to what they expect or want from a situation and what is important to them.

> **Key vocabulary**
>
> *Pragmatic* – concerned with the practical and contextual. Pragmatic communication provides all anticipated context to its audience to facilitate and enhance understanding.

Understanding who your audience is and having a clear understanding of your purpose are especially important in IT communications because they can enable the following key attributes of effective technical communication:

1  Meeting your audience's needs
   Technical communication is produced for a wide variety of audience types. These can include people who have a background in IT as well as those who don't. This background has an impact on how you communicate, including whether you do or don't use technical vocabulary, and whether you need to explain what a concept is or if you can assume your audience is already familiar with it. Some audiences will rely heavily on visual cues like diagrams, screenshots, and illustrations to understand your message, while others may find visuals unnecessary. Some members of your audience may not use English as their first language and others may. A careful analysis of your audience with these considerations will make your communication more likely to match their needs, making your message more effective.

2   Ensuring messages are concise

In technical communication, readers want to get in and out of a message as quickly as possible. Being concise (or the noun form – concision) is a key attribute, so knowing who you are creating content for and why will help you to avoid including unnecessary details, explanations, or data that take up too much of your reader's time and hide your main point. Remember that a concise message should be accessible. Make sure to structure your message effectively so that your reader doesn't have to spend unnecessary time or effort finding the information that is important or useful to them. Your readers should be able to read both vertically (by scanning down a text) to find the information they need, as well as horizontally (across the page in wordy paragraphs like the one you are reading right now). Making sure your messages are quick and easy to scan and read helps to ensure your communication is pragmatic. This is also especially true when readers come with different linguistic and cultural backgrounds.

3   Meeting your communicative targets

Beginning with a clear purpose in mind can help to ensure that you meet your objectives. Knowing the overall purpose of your message, whether you want to provide or request information, to persuade someone of a course of action, or to collaborate with a team on an extended project, keeping your purpose firmly in mind will help you to make sure that your messaging is appropriate for purpose. This includes language features, including style, tone, and register. For example, a technical report intended for management may focus on summarizing the key findings and implications, while a communication to a development team should usually delve more deeply into the technical details.

4   Tailoring your approach

Knowing your audience also means knowing their role and responsibilities within the specific context and being aware of their associated interests. Knowing their role can help you to highlight the details and concerns that are likely to be most important for them. Executives are likely to be concerned with big-picture takeaways and summaries of how a product, proposal, or policy may impact the whole organization. Software and product developers need a focus on more minute (specific) details and design specifications. However, executives from different contexts or developers who work for different companies may also have different expectations of what and how you communicate.

5   Developing credibility (ethos)

Being able to communicate effectively to a target audience with a very clear purpose in mind gives benefits to your role within the organization as it builds trust in your work and credibility in your opinions with your audience. When you demonstrate that you understand your audience's needs and desires, you enhance your own reputation with your audience and demonstrate professionalism and expertise. For more information about the importance of ethos in technical communication, see Chapter 3.

---

### Key vocabulary

*Register* – this refers to language choices that are made to reflect the connection between the message and its audience. See formality, later in the chapter for more details.

*Style* – this refers to the broader language choices made to convey the message. Sentence structure, for example, using passive or active voice, is one example of how to determine style.

*Tone* – this refers to the general feeling creating in a communication. Positive/negative language, and the use of adjectives are some examples of ways to influence tone.

## 1.2   Analyzing audience: who's going to read this?

Three considerations can help you start to analyze your audience so that you can better achieve your communication targets.

### What is their connection to the event or situation?

The first question asks you to consider your audience's relationship to the communicative event, such as their role in a project, their level of involvement, and their expectations. Do they need to act? Do they need to only be aware of a situation? Are they internally connected or are they external observers who are not part of the organization?

---

**Example: work presentation**

If you are giving a presentation to a group of colleagues, you will need to consider their role(s) in the company, their level of knowledge about the topic, and their expectations for the presentation. You may need to account for differing language or cultural backgrounds. Very often, these roles, the knowledge, and the expectations will also include variation. This means you need to both be aware of this variation, and you also need to account for it.

---

### What is their background in IT?

The second question considers your audience's IT background, which includes their level of knowledge about IT, their experience with using IT, and their attitudes towards IT. We have already mentioned that not all technical content consumers have education or experience in IT. Your audience's knowledge and experience with technology will impact how they read a technical document. Even when your audience has a technical background, it is important to remember that IT is a very broad area, so a user with a lot of experience in online gaming will not necessarily understand data modelling. When considering your audience's likely background knowledge of IT, you should consider how likely they are to be familiar with parallel or closely related situations or tasks.

---

**Example: technology roadmap**

If you are writing a technical document like a technology roadmap for a group of IT engineers, you can assume an understanding of technical terminology and use it freely. Theoretical frameworks can be considered common knowledge. On the other hand, if you are creating instructions for updating an application, you will need to use more general language that all users will understand, avoiding technical terminology or jargon.

---

### What do they need from the text?

The third question considers the audience's pragmatic needs from the text, such as what they need to know, what they need to do, and how they need to feel. You might need to inform the manager of a different department of planned service outages so that they can warn their team not to begin an important task on company systems at the wrong time. In this case, you would need to inform them of the time, duration, and services interrupted. You would need to advise them about how to keep hardware and software safe and data secure. You would need to reassure them of the importance and need for the planned outage, so that any possible feelings of inconvenience are minimized.

---

**Example: user manual**

If you are writing a user manual for a new software application, you will need to make sure that the manual provides the user with the information they need to use the software effectively. Considering the first point: users need to know that they have the necessary system requirements and the tools necessary. They need to follow the process outlined in your procedure in the correct order. They need to feel confident that they are following these steps correctly and that they are completing the procedure as directed. Effective user manuals achieve this by providing bulleted lists of requirements, minimizing language use, using numbered steps, and including clear visuals that support the actions of the user.

---

By considering these three questions, you can gain a better understanding of your audience and tailor your communication to their needs and interests. This will help you to improve the effectiveness of your communication.

## Imbalances

We have already mentioned that not every audience of a technical document has education or work experience in IT. This creates imbalances in use, context, and knowledge of IT information. Imbalances can lead to communication barriers because what one individual knows and expects is different from what the other knows and expects. Imbalances can be related to explicit and obvious differences like unknown terminology or differences in usage. They can also be related to implicit challenges, like unstated consequences or results of a choice or action. Someone with more IT knowledge may assume that their reader will automatically understand the consequences of a choice, but if the reader does not share the same level of knowledge, this is unlikely to be true.

Consider an API (Application Programming Interface) like a Google account. Many individuals and businesses hold Google accounts, but they are used very differently, depending on who is using them and why. An individual frequently uses a Gmail account for personal email and for signing up to and receiving alerts from accounts that require the use of email. Some individuals use their Google account to sign up for other services on third-party sites and apps instead of creating new user login details for the third-party site.

Businesses also use Google's services, especially small to medium-sized firms. They often use a domain name that is branded to their organization but is linked to a Gmail account and Google services. In that case, professionals are using Google's products to complete work-related tasks.

The small- to medium-sized companies that use Google's products often also have IT technicians and staff to support technology-related aspects of their business. These people would have a different working purpose and understanding of these products' capabilities, limitations, and integrations with their company work.

Finally, there are the back-end employees of Google itself who offer technical and digital security support to all three. Those Google employees are likely to have very different knowledge, use of the products, and contexts.

What does this mean for writers of technical documents? You need to carefully consider the knowledge, use, and contexts of use of your intended audience. You need to be clear about everything and not assume that your audience will automatically understand any unsaid results or consequences. You need to consider what will matter most to them and what does not need to be communicated.

## 1.3   Common types of IT audience

The nature of IT communication means that audience types vary considerably and almost everyone counts as an IT audience at some point. We know that not everybody has the same level of experience and knowledge around IT tools and systems. Different groups tend to have different levels of experience. Different audiences also consume technical content and read technical texts for different reasons. Knowing which audience you are addressing can help you to consider your specific audience's needs.

Generally, there are three categories that are considered most common:

**Technical experts**: people who have a deep understanding of IT concepts and/or architectures. These people work in IT and/or have studied it extensively. Technical experts are usually involved in the design, development, maintenance, or implementation of IT systems and products. Think of the Google employees in the earlier example. Their understanding of the technology used by their API is likely to be the deepest of all groups.

**Business users**: these are people who use IT products and services as part of their job. Sometimes lay managers may be involved in the oversight of tech companies, without having a professional or educational background in IT. They may have some understanding of IT, particularly the systems and products that they use as part of their employment, but they are usually not experts and often require troubleshooting support. Think of the business users who may have their own domain but use Gmail and other Google products to operate their business. Some of the employees may have an IT background, but others do not and use the products to do their job.

**General users**: these are people who use IT products and services in their everyday (domestic) lives. They do not necessarily have training or any interest in IT beyond personal entertainment and communication purposes. An example could be individuals with personal email and cloud storage accounts with Google. They may have a background in technology, or they may use the products because they are very common and free.

How you communicate with each of these audiences will be different because you will need to tailor your communication to their needs and level of understanding.

## 1.4   Analyzing purpose: why are they reading it?

Three common types of purpose in IT communication are informative, instructional, and persuasive. Each purpose has distinct implications for the communication process. Here's an explanation of each purpose and its impact on the decisions you as a technical writer will need to make:

**Informative purpose**: when the purpose of IT communication is to inform, your primary goal is to provide factual and relevant information to the audience. This purpose is commonly used in situations where your audience needs a better understanding of a particular topic, concept, or update. Informative communication focuses on delivering accurate and up-to-date information in a clear and concise manner. It helps to enhance knowledge, raise awareness, or provide updates on IT-related matters.

---

**Example: project documents**

An example of an informational IT-related document is a project document. A project document is meant to keep teams aligned to the same goal. The effect of informative communication is to educate the audience, increase their understanding, and foster informed decision-making. Employees read project documents to understand the thinking behind design choices or to know about next steps or constraints.

---

**Instructional purpose**: instructional communication in IT aims to guide or direct the audience on how to perform a specific task, procedure, or operation. The purpose is to provide step-by-step instructions, guidelines, or tutorials to help the audience achieve a desired outcome or execute a particular process effectively. Instructional communication often includes detailed explanations, demonstrations, and examples to facilitate learning and comprehension. Instructional IT-related content enables the audience to acquire new skills, enhance their capabilities, and successfully carry out specific IT-related tasks.

---

**Example: legacy documentation**

Legacy documentation is written for individuals who will be working on later versions of software. It is created for the future coders of that programme, to help them understand the design choices and thinking around why a product has been put together in a particular way. Legacy documentation enables those future coders to work with a greater understanding of the decision process that informed the original product. It helps them to do their job more effectively because they are more informed.

---

**Persuasive purpose**: when the purpose of IT communication is persuasive, the goal is to influence the audience's attitudes, beliefs, or behaviours regarding a particular topic or course of action. Persuasive communication aims to convince the audience to adopt a specific viewpoint, support a particular idea, or take a desired action. For example, a persuasive internal communication may encourage employees to adopt two-factor authentication practices to maintain higher data security. This type of communication employs persuasive techniques such as logical reasoning, emotional appeal, evidence, and argumentation to persuade the audience to align with the communicator's position or recommendation. The effect of persuasive communication is to motivate the audience to make decisions, change their behaviours, or support a specific IT-related initiative.

---

**Example: international IT research conference invitations**

International conference invitations are an example of persuasive communication in IT. For example, the 3rd International Conference on Power Electronics, Smart Grid, and Renewable Energy (PESGRE 2023) invites papers from researchers studying within the area of power and renewable energy [7]. The conference offers the opportunity to publish and provides presenters with the option of giving their presentation remotely. Both are persuasive appeals aimed at convincing researchers to participate in the conference. The first appeal is aimed at researchers who need to publish original research in their field as part of their job. The second appeal is aimed at researchers who may live far from the presentation or may not have the money or funding to pay for travel expenses. These benefits may not be obvious to readers who are not familiar with the contextual expectations of research, publication, and collaboration in research and academic circles, but the intended audience of the conference invitation are likely to have these concerns. That makes this appeal an effective example of using persuasive strategies effectively. This conference invitation considers the needs, wants, and challenges that its intended audience is likely to face and uses appropriate strategies to convince researchers to participate.

Each type of purpose affects communication in distinct ways:

- **Informative communication** focuses on clarity, accuracy, and the delivery of relevant information. It emphasizes the transmission of knowledge and understanding. This is knowledge and understanding from the perspective of the audience consuming the content. It is fact-based, rather than opinion oriented. The reader should end with a clearer understanding of the topic. A sign outlining what can and cannot be taken into a server room should clearly deliver its message and leave no room for confusion.
- **Instructional communication** emphasizes the provision of clear, actionable guidance. It aims to facilitate learning and skill development. The reader should end with a completed task and/or knowledge of how to complete a given task in future scenarios. The sequence should be clear, and the user should be able to follow along to complete it.
- **Persuasive communication** emphasizes persuasion techniques to influence attitudes, beliefs, and actions. It seeks to bring about a change in behaviour or garner support for a particular viewpoint or course of action. The reader should be convinced of the benefits of adopting the strategy or viewpoint or be encouraged to act in the way that the communication supports.

### Persuasion in technical communication

Just like in other types of communication, persuasive technical communication uses different types of strategy to convince its audience. Rhetorical studies originated with the ancient Greek philosopher Aristotle. Aristotle believed there were three main strategies for convincing your audience: to use *logos* (or reason – facts, statistics, knowledge, or data), *ethos* (credentials and expertise), or *pathos* (appeals to emotion or shared values). These strategies help to convince your audience of the rightness of your opinion. Knowing your audience helps you to better choose the best persuasive strategies to help convince them.

In technical communication, two common persuasive strategies are ethos – relying on credentials for authority and pathos – relying on data for authority. Relying on credentials for authority happens when the expertise and position of the writer are used to convince the audience of the main message. When facts and statistics are used to persuade the audience, this is relying on data for authority.

While pathos is used sometimes in technical communication, it is less common. Pathos, or appeals to the emotions and/or shared values of the audience, is commonly found in the

**Key vocabulary**

*Ethos* – persuasive strategies related to the credentials or trustworthiness of the speaker or source of information.

*Logos* – persuasive strategies that use facts, statistics, or data to convince the audience.

*Pathos* – persuasive strategies that appeal to the audience's emotions or values.

rhetoric used to market digital security products and to warn users and employees against the hazards of online data management. In workplaces, it is also used to convince employees to follow guidelines and adhere to best practices.

These three persuasive strategies are outlined with a following example.

### Relying on credentials as authority: white papers

An example of this type of persuasive communication is a white paper. A white paper is a reliable document or guide that usually tackles complicated topics and challenges. Its purpose is to teach readers and help them make well-informed choices. White papers are usually written by subject matter experts with deep knowledge in the field or industry that is the focus of the paper. It is the credentials (academic, professional, or experiential) that provide the authority for the white paper. Readers rely on this knowledge to gain a deeper understanding of the topic.

Table I gives some examples of the type of expert that often writes a white paper:

### Relying on data as authority: scientific research articles

Another type of authority can come from presenting facts and data as evidence. Scientific research articles are an example of this. They present or test theories and experiments that rely on what has already been established as known within that field. In order to be convincing, the writers of a scientific journal article need to present sufficient information about the research design and data collection methods for readers to decide how valuable and/or accurate the research study is to the field.

### Relying on shared values and emotional appeals as authority: cybersecurity marketing materials

The third, less common type of authority comes from appeals to common ground or shared emotions or values. Corporate Social Responsibility (CSR) Reports are one example where the

*Table I* Examples of white paper writers

| | |
|---|---|
| **Researchers** | Professionals who conduct research in their field and share the results of this research |
| **Industry specialists** | Working professionals who are recognized for their experience and expertise in a given field |
| **Academics** | Professors and scholars who conduct research and studies to further a specific area of knowledge |
| **Technical experts** | Scientists and engineers who can explain complex technical concepts and innovations |

authority of the message comes from an appeal to shared values and sometimes to emotions. Look at the following statement about corporate social responsibility from Microsoft's website [8]. What shared values are suggested?

Expand opportunity
We believe economic growth and opportunity must reach every person, organization, community, and country. This starts with ensuring everyone has the skills to thrive in a digital, AI-enabled economy, and extends to empowering nonprofits, entrepreneurs, and other organizations to digitally transform and address society's biggest challenges. [8]

The aforementioned shared values can be seen in the appeal to each person's opportunity to develop digital and AI-enabled skills. It also appeals to nonprofits and the greater good of society. It is implied that the development of AI tools by this company will not be at the expense of specific communities and groups.

By understanding the specific purpose of the IT content you are creating, you can tailor your IT communication messages, structuring their content, and selecting appropriate techniques to effectively achieve your desired outcomes. Whether the aim is to inform, instruct, or persuade, aligning the communication approach with your specified purpose enhances your message's impact and increases the likelihood that you will achieve your goals.

### Strategies to identify purpose

As you become more experienced in creating IT messages, you will develop the skills to quickly identify your purpose. Identifying the purpose of IT communication involves analyzing the content, context, and intended outcomes of your communication. Here are some techniques and indicators that can help you be clear about your purpose:

**Consider your audience's response** – what do you expect/need the audience to do as a result of your communication? Will your communication teach them how to complete a task or install a programme? Will your communication notify employees of updated protocols or guidelines? Do you want your audience to start doing something or stop doing something? Do you want your audience to agree to fund a project or select a service?

---

**Example: opinion pieces**

Opinion pieces from thought leaders are an example of persuasive IT communication that aims to encourage more people to use a product or service. Thought leaders are recognized experts in an industry. They may be invited to interview on a podcast or share their views in an article. The aim is to influence industry trends by getting a wider public to follow their advice, buy a product, adopt a habit, or stop doing something.

According to Roger's theory of diffusion [9], an innovation follows a predictable path in becoming mainstream or popular with, known, and used by a lot of the general public. It first gains popularity with early adopters (people who are quick to try new products or experiments). Next it becomes popular with the early majority before gaining users in the late majority and laggard (people who are very slow to make the change) categories. Smartphone usage demonstrates this: Apple sold over 1.4 million units of its iPhone 1 mobile phone in 2007 [10]. By 2015, they had sold over 230 million units of their smartphone [10]. Buyers in

> 2007 represent early adopters, with people in 2008–2010 representing the early majority. By the time the iPhone is popular globally, it has moved into late majority territory.

**Analyze the language you have used in a first draft or in a sample you find from the internet** – what are the key verbs used? Which of the three main purposes do they best match? Table II gives some examples of verbs that are commonly used for each purpose.

**Consider the consequences of failing to communicate your message** – what happens if you don't send your message, or your message is not properly received? Are there legal or financial consequences of your audience missing your message? The example that follows demonstrates one real-world case of a poorly communicated message.

**The consequences of failing to communicate: NASA PowerPoint and Columbia accident**

There is a well-known example of the serious consequences of poor technical communication in the United States space travel history. In 2003, seven astronauts aboard the Columbia space shuttle died when large chunks of the foam insulation on the external tank of their space shuttle broke off and hit the left wing of the shuttle. While the shuttle orbited in space, scientists at NASA debated how to fix the shuttle and whether it could safely re-enter the Earth's atmosphere. They decided re-entry was a safe option. Later, when the shuttle attempted re-entry into Earth's atmosphere, all seven individuals on board lost their lives when the craft imploded.

Officials had been briefed by NASA experts with a PowerPoint slide deck that included information warning of the potential danger. However, the information was so poorly organized and worded that the warning was completely missed.

A government investigation later revealed that this was part of a larger organizational problem at NASA of using slide decks instead of writing technical reports [6]. This bad habit of using an inappropriate tool and poor messaging resulted in the deaths of seven astronauts in February 2003. Clearly, being able to communicate clearly is a vital part of technical communication [11].

*Table II* Different verbs for most common purposes

| Inform | Instruct | Persuade |
| --- | --- | --- |
| - provide | - show | - convince |
| - give | - demonstrate | - argue |
| - explain | - learn | - encourage |
| - describe | - outline | - urge |
| - define | - execute | - mobilize |
| - illustrate | - teach | |
| - present | | |

## 1.5    Language for audience and purpose

Here are some specific language guidelines regarding each of the main technical audience types. Table III compares the precision of language, use of visuals, and content included according to each reader's profile. More explanation of how visuals are used in technical communication is provided in Chapter 2.

### Formality

Formality is an example of how language changes according to audience and purpose. Remember the term 'Register' from earlier in the chapter? Formality is a way of using language appropriately according to the social context. We use more formal language for more official and serious situations. We use more informal language with people we are closer to and when our connection is casual and/or friendly. An example that is common for many people is the difference in how we communicate with our siblings (brothers and sisters) and how we communicate with older family members like grandparents. Frequently, people use more informal communication with brothers and sisters and are more formal with older relatives. The formality is used as a way of showing respect.

Generally in the workplace, formality is a way of acknowledging the power distance and relative ranking of individuals and/or groups and the seriousness of a situation. Usually, the closer two individuals or positions/offices are, the less formal the communication.

We can see examples of formality in other areas of life. In the military, for example, subordinates (soldiers or cadets with a lower ranking) are required to salute and stand at attention when a superior (an officer with a higher ranking) arrives. The salute is a way of demonstrating respect for their higher position within the military structure. It also demonstrates that the cadet recognizes the hierarchy of the military and their position within it.

Another place we can see formality used is in correspondence – emails, letters, and memos. An email should start with a salutation or greeting, which identifies the intended reader (see Chapter 3 for more information about workplace communications). The salutation shows how formal the email is and how close the relationship is between the sender and receiver.

*Table III* Types of technical audience and language choices

|  | Technical experts | Business users | General/non-technical readers |
| --- | --- | --- | --- |
| **Precision of Language** | Use technical language Define any potentially unfamiliar terms | Use plain language Avoid technical language and jargon. | Use very plain language Avoid any technical language or jargon. |
| **Use of Visuals** | Diagrams used to exemplify data flow and processing | Graphics to illustrate data and processes | Images to illustrate a process or configuration |
| **Content** | Comprehensive: covers a wide range of theory and practical information | Emphasis on data and practical usage | Emphasis on practical use and general background |

*Table IV* Examples of different purposes

| Context | Example | Analysis |
|---|---|---|
| **Texting to friends** | RU coming 2day? | • shortened forms<br>• use of slang<br>• use of numbers and letters to substitute for words<br>• use of nicknames |
| **Completing an incident report** | The attempted ransomware attack was identified at 0235 23/02/2023 by Security Analyst Femi Obelayo. | • passive voice<br>• full-time and date structures<br>• no contractions or shortened forms<br>• complete legal names and titles |
| **Write a safe password policy** | Our company requires the use of strong passwords. Strong passwords must be a minimum of 16 characters long and contain at least one of each of the following:<br>• uppercase letter<br>• lower case letter<br>• number<br>• symbol<br>In addition, passwords cannot be made up of any dictionary-based words.<br>No more than two consecutive letters or numbers can be used. | • formal language<br>• nominalization (see Chapter 5 for more on grammatical metaphor)<br>• precise language<br>• bulleted lists to detail requirements<br>• passive voice |

Consider the difference in formality in the following salutation examples in English:

Greetings
Good Afternoon
Hello
Hi
Hey
Dear Ms. Mathew,

*Formality and purpose*

Table IV contrasts different purposes and some of their language features.

## 1.6   Review

The following questions will help you consolidate your knowledge about technical audiences and purposes. They will help to deepen your understanding through practice.

## Reflection questions

1 What types of audience do you usually write for in your current capacity?
2 Why is technical communication considered pragmatic?
3 When have you encountered difficulties as a reader or a writer?
4 Who are technical writers communicating to?
5 What are some of the common interactional exchanges where you live? What purpose do they serve?
6 How confident are you in your current ability to modify your writing for different audiences and purposes?
7 What are some of the different ways that formality is expressed in your context/language?

## Application tasks

1 Choose a common form of communication you use regularly. Analyze your audience using the following questions:

1 Which of the following terms would you use to describe your audience's level of expertise?

a) Advanced technical knowledge
b) Intermediate technical knowledge
c) Basic technical knowledge
d) Non-technical background

2 What is the primary goal of your audience when interacting with your content?

a) To understand complex technical concepts
b) To make informed business decisions
c) To gain a general understanding of the topic

3 Which of the following writing styles would best suit your audience?

a) Technical jargon and detailed explanations
b) Clear and concise language with minimal technical terms
c) Engaging storytelling with relatable examples

4 How familiar is your audience with the industry-specific terminology?

a) Very familiar, they use it regularly
b) Moderately familiar, they encounter it occasionally
c) Not familiar, they are new to the industry

5 How would you describe the level of detail your audience expects?

a) In-depth technical information with step-by-step instructions
b) High-level overviews and key points
c) General concepts with practical applications

6 What is the preferred format of your audience for receiving information?

a) Detailed technical reports or whitepapers
b) Summarized data and charts
c) Plain language explanations and examples

7   How important is it for your audience to see real-world business examples in your content?

   a) Essential, they need concrete examples to understand the context
   b) Useful, but not crucial for their understanding
   c) Not important, they prefer theoretical explanations

8   How patient is your audience with complex or technical explanations?

   a) Highly patient, they can handle complex information
   b) Moderately patient, they prefer simpler explanations
   c) Not patient, they want straightforward and easy-to-understand content

9   Which of the following types of visuals would be most helpful for your audience?

   a) Detailed diagrams and technical illustrations
   b) Charts and graphs showcasing relevant data
   c) Engaging images and infographics

10   How familiar is your audience with your specific product or service?

   a) They are experts and use it extensively
   b) They have a good understanding but are not experts
   c) They are completely new to your product or service

**2**   Complete the audience analysis using the scenario that follows.
*Scenario:*

> *You have been assigned to a working group tasked with organizing a workshop on best practices for maintaining cybersecurity in the workplace. The objective of the workshop is to educate and raise awareness among fellow employees about the importance of cybersecurity in today's digital landscape.*

Your working group needs to analyze the audience to ensure that your workshop effectively addresses the needs and interests of the employees who participate.

**Relationship to communicative event: participants in a cybersecurity workshop**

- What roles will the workshop participants have in your organization?
- Do participants have the choice to attend or is attendance mandatory?
- How does the answer to question 2 affect how your workshop is structured?
- What will employees expect to do in the workshop?

**IT background**

- How much do your fellow colleagues likely already know about cybersecurity?
- Have employees had any prior training?
- Are there pre-existing structures that employees are familiar with?
- Do employees know the terminology you will use?

**Pragmatic needs**

- What do participants in the workshop need to know after it ends?
- What do participants need to do (or not do) after the workshop ends?
- What are the responsibilities and expectations in your organization for individual employee cybersecurity?
- What are the consequences of non-compliance?

**3** Conduct a web search to find a sample user guide for a product or service.

    a) Which of the three most common purposes in IT does this manual serve? (see Section 1.4 for more information)

    b) What are some specific features that suggest this?

    c) What are some specific language examples that support this?

    d) Does your sample achieve its purpose? Why/not?

**4** Use the scenario that follows to conduct another audience and purpose analysis:

*Scenario:*

    *You are a technical writer working for a medium-sized web application design company. Your company is primarily concerned with creating API applications for small businesses to automate and streamline their processes. You have been tasked with creating a user manual for a new software product that the company is launching. The manual should be written in plain language and should be easy to understand for a reader who doesn't have an IT background. Your target audience for the app are small business owners who are not familiar with web and mobile technical terminology. Complete* Table V.

**5** Conduct a search for a white paper online. Use the following questions to analyze it.

    **5a** What kind of persuasion is used to write this white paper? Why? What can you find out about the authors? Who commissioned the report? What basis does that give us to trust the information?

    **5b** What kind of audience does this appear to be written for? Identify three examples from the text as evidence.

    **5c** What examples of formality can you draw from the text? What does that suggest about purpose and audience?

**6** Analyze the example following email.

Subject: urgent: ransomware attack incident response

Dear IT Team,

I regret to inform you that our company has recently experienced a ransomware attack. The attack breached our email firewalls through an attachment that was opened. I want to assure you that we are taking immediate action to address this situation.

*Table V* Audience analysis prompts

| Prompt | Analysis |
|---|---|
| What is your target audience's relationship to the manual? | |
| What is your target audience's background in IT? | |
| What does your target audience need from the app? | |
| How can your manual help to meet your target audience's needs? | |
| What does your target audience want from the app? | |
| How can your manual help to meet your target audience's wants? | |
| What is the main purpose of the manual? | |
| Are there any other purposes that the manual might be used for? | |
| How formal should the language in the manual be? Why? | |

Please read the remainder of this message (Table VI) carefully to understand your role and responsibility during this time.

*Table VI* Immediate next steps

| Isolation | Security analysts quickly identified the attack and isolated affected systems. Network administrators immediately quarantined the affected network. |
|---|---|
| External support | A professional cybersecurity client: Cyber Solutions has been contracted to help us assess and move forward. |
| Data backup | Backup specialists have begun the process of restoring backup data systems. |
| Security audit | Security analysts will work in tandem with Cyber Solutions consultants to evaluate the current system vulnerabilities and assess measures to strengthen our security. |
| Communication | Communications with all employees and clientele are being managed through the Communications Manager: Huda Al Raisi. Her team will be liaising with department heads on next steps. |

**Client notification:**
Please understand the severity of this situation and **avoid disclosing the event to any external parties**. At this time, we have not yet notified clients about the incident. We are working diligently to gather all relevant information and assess the scope of the attack. Once we have a clearer understanding of the situation and its potential impact on our clients, we will make an informed decision regarding client notification.

**Confidentiality:**
I want to emphasize the importance of maintaining strict confidentiality regarding this incident. Please refrain from discussing it with anyone outside of the IT department and report any suspicious activity or information immediately to the incident response team.

This is a challenging time for our company, but we are committed to resolving this issue swiftly and minimizing any potential impact on our clients and business operations. Your dedication and expertise are critical in helping us through this crisis, and I appreciate your hard work.

We will continue to provide updates as we gather more information. If you have any questions or concerns, please do not hesitate to reach out to the incident response team or me directly.

Thank you for your understanding and cooperation.
Sincerely,

Olayuke Tunde
IT Manager
Assertco

6a  Identify the intended audience. What type of audience are they? How do you know?
6b  What is the relationship between the writer and the reader? What specific language demonstrates this?
6c  Find instances of appropriate formality for this context.
6d  The email is long. Are there any language changes that could be made? If so, do they impact formality, purpose, or audience?

7  Conduct a search for YouTube's 'Fair Use' policy. Use the keywords 'YouTube + fair use + policy'.

    **7a**  Choose a paragraph to analyze.
    **7b**  What type of audience does the paragraph seem to be written for? Why?
    **7c**  What does the paragraph suggest that the reader needs to know?
    **7d**  Look at the word choice. What does this suggest about the intended reader's knowledge?
    **7e**  Look at grammar and sentence length. What do they suggest about the intended reader's reading skills?

8  Text Analysis: review the following text

Dear Team,

I trust this email finds you well. I am writing to remind everyone of the importance of maintaining our data security at this company.

    Every day, we handle sensitive information that forms the backbone of our operations. This information includes our clients' data and information that helps our operations to succeed. Our reputation is contingent upon maintaining the integrity of this data. When we talk about data, we are talking about our collective efforts and our professional integrity – as individuals and as a company.

    Recently, our IT department has updated our data protection policies to better defend against cyberattacks and data leaks. They can be found on the IT Portal of our company intranet, and I have attached a copy to this message for your convenience. You are kindly requested to read through them carefully so that you are fully aware of how our company is protecting client and operational data.

    I urge each valued member of our team to diligently protect this data and our collective reputation by diligently adhering to our updated security guidelines. These guidelines protect our clients and our reputation and the trust of our clients.

With sincere appreciation for your continued diligence and efforts.

Paul S. Almeida
CTO

    **8a**  What is the purpose?
    **8b**  What persuasive strategies have been used?
    **8c**  Who is the intended audience?
    **8d**  What examples of formality do you see?

9 Review the following communication and answer the application tasks that follow.

## MEMORANDUM

Subject: mandatory compliance with enhanced data security protocols
To: all employees
From: Diane Paula, Chief Technology Officer

This communication serves as an official reminder of the expectation that all employees fully adhere to the company's security protocols. Global trends over the past several years show the significant rise in cybersecurity attempts and the threat they pose to business operations.

Data breaches not only compromise company assets, but they also pose the potential to endanger client and employee personal information. Finally, they pose a serious legal and financial risk, together with the negative impact on corporate reputation. Given the severity of these consequences, it is imperative that all employees understand the severe consequences of data breaches.

The IT department has mandated strict compliance with the recently updated security guidelines. Key aspects of the security guidelines include

- Regular password updates and the use of strong, unique passwords
- Access of sensitive information limited to secure and authorized channels
- Immediate reporting of any suspicious activities or potential security breaches

Full compliance with all of these guidelines is the responsibility of each employee. Your cooperation and diligence in following these protocols are crucial to the ongoing security and integrity of our digital infrastructure, personal information, and corporate reputation. Failure to comply with these guidelines will be subject to review and may lead to disciplinary action.

Thank you for your continued support and commitment to safeguarding Safetyco's digital assets.

**9a**  What is the purpose?
**9b**  What persuasive strategies have been used?
**9c**  Who is the intended audience?
**9d**  What examples of formality do you see?

## Works cited

1  "Definition of technical communication," *Technical Communication Body of Knowledge (TCBoK)*, 2023. [Online]. Available: www.tcbok.org/about-the-technical-communication-body-of-knowledge/definition-of-technical-communication/ [Accessed 10 April 2024].
2  C. E. Shannon, "A mathematical theory of communication," *The Bell System Technical Journal*, vol. 27, no. 3, pp. 379–423, 1948.
3  W. Schramm, "How communication works," in *The process and effects of mass communication*. University of Illinois Press, 1954, pp. 3–26.
4  E. Goffman, *Interaction ritual: Essays in face-to-face behavior*. Routledge, 1967.
5  P. Watzlawick, J. H. Beavin, and D. D. Jackson, *Pragmatics of human communication: A study of interactional patterns, pathologies and paradoxes*. W. W. Norton & Company, 1967.
6  R. Scollon and S. W. Scollon, *Intercultural communication: A discourse approach*, 2nd ed. Blackwell Publishers, 2001.
7  "PESGRE 2023 – PESGRE 2023." Available: https://pesgre2023.org/
8  "Microsoft Corporate Social Responsibility | Microsoft CSR." Available: www.microsoft.com/en-us/corporate-responsibility
9  E. M. Rogers, *Diffusion of innovations*, 5th ed. Free Press, 2003.
10  "Annual sales of Apple's iPhone (2007–2021)," *GlobalData*. Available: www.globaldata.com/data-insights/technology – media-and-telecom/annual-sales-of-apples-iphone/#:~:text=Apple%20sold%20over%201.4%20million,of%20%24630%20million%2C%20in%202007
11  "NASA | Columbia Accident Investigation Board." Available: https://history.nasa.gov/columbia/CAIB.html

Chapter 2

# Design principles in IT

## 2.1 Introduction to design principles

This section introduces the key principles of design in technical documentation, explaining why design matters in technical documents.

### Why are design principles important in technical communication?

Design is important in any professional documentation, and this is equally true in technical documentation. There is an extensive field of research that looks at how design can enhance the usability and user experience of technology [1]. The Nielsen Norman Group defines usability as how easy and pleasant an application's features are to use [1]. The same is true for more traditional technical documents like emails, reports, or instruction manuals. Good design helps the reader to navigate the text more easily and to better understand how information is related. It also helps your reader trust you, since good design will make your technical texts seem more professional [2].

Design can give the reader extralinguistic (non-text-based) clues about how to understand a process or concept and can provide a visual anchor. This means that even if you don't understand the language a text is written in, you can identify textual elements like titles, headings, and images with their captions. How those parts have been put together gives you, the reader, information about how to connect and interpret them [3]. This is because of how our mind and eye reads information visually. Most people are visually oriented rather than data oriented. Putting information together mindfully on a page or screen will help your reader to better digest that information [4]. Design can guide the reader's eye and attention to what is most important. Design helps the reader not to miss important information and also to understand the connection between parts on a page or across a document [5].

### Skimmable takeaways and easy access points

One of the most important points to consider when designing an IT document is why your reader is engaging with your content and how they will be engaging with it [6]. As discussed in Chapter 1, technical documents are read to achieve a purpose, and sometimes that means that they are read in a vertical (down the page) rather than horizontal (across the page) manner. It is extremely important to write content that can be skimmed quickly and facilitates vertical as well as horizontal reading. In the case of technical documents, where people are already reading to quickly find what they need and get out again, this becomes especially true.

DOI: 10.4324/9781032647524-2

Your document can be made more skimmable if you use effective design. This is because the reader will be able to identify the patterns that help them to quickly navigate your text.

## 2.2   Design principles

Effective documents consider more than the message as conveyed by words. Effective design can be seen even if you are not able to read the language of the communication. Together, the design principles work to guide you effectively through a document or multimedia. This is based on the principles of Gestalt theory [3].

In order to discuss the design principles effectively, it is important to have some shared language. A term that will be used in this section to talk about design is 'element'. An element can be text, a visual, or a heading. It could also be a border. An element is something on the page. Some elements naturally attract our attention more than others, so we sometimes talk about elements as having more weight than others. When an element is described as having more weight or weighing more than other elements, we mean that the eye is naturally drawn to it more than other elements.

The list of design principles is

- Unity
- Balance
- Alignment
- Hierarchy
- Emphasis
- Proportion
- White Space
- Repetition
- Movement
- Contrast

### Unity

Unity is how well the different elements and design choices you make work together. It means tying things together so that they go well. Colour is a good example of how unity can be effective or ineffective. When colours are used effectively, they show which elements are part of the same order and which should be emphasized for difference. When colour is NOT used effectively, elements can blend into the background or each other.

Another example of unity can be found in shape and size. Using shapes in a consistent manner can build unity throughout a document. Size, too, can help ensure that proportion creates a sense of cohesion.

The elements in Figure 2.1 provide a sense of unity that is created through the use of colour and shade repetition. Shape, size, and black and white have all been used effectively to create a sense of unity to this image. When unity is used effectively, the reader can better follow your content, since they are led through the use of patterns, colours, and repetition to recognize and understand the elements on a page. Using a simple grayscale example, we can immediately recognize how the elements on this page contribute to a sense of wholeness.

In Figure 2.2, on the other hand, there is no clear pattern to follow. The design does not represent unity. It is harder to recognize how the parts are connected. The white shape is hard to distinguish, so it could be read as an absence or as a white shape. This image feels unfinished, inaccurate, or incomplete. This is the result of a lack of visual unity.

## Balance

The term balance is used to refer to how much an element draws your attention relative to other elements on the page or screen. It is also used to talk about using the overall area. Good balance means using the whole space (page, screen, etc.) effectively, and not crowding content into one part. Good balance also recognizes that the eye is naturally drawn to some features more than others [7]. Images, for example, attract more attention than paragraphs of text. Borders are more attractive to the eye than text, and colours draw the eye more than black and white. Placing elements carefully on the page in order to avoid the danger that your readers miss important content involves understanding which elements draw the eye and which do not – and then placing them

Figure 2.1 Example of effective unity

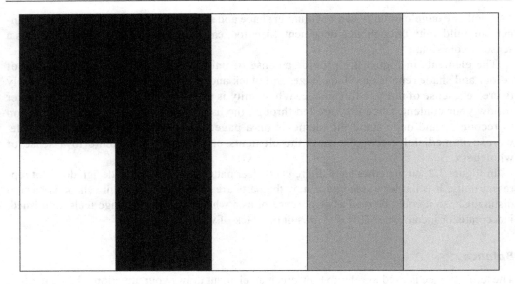

*Figure 2.2* Example of ineffective unity

on the page or screen accordingly to where the reader is likely to notice them. This is strongly connected to movement, which is discussed later in this section. Balance is also closely connected to white space.

The example in Figure 2.3 is a good use of the screen or the page. Elements are positioned in a way that is helpful to a reader. Whether the language is English, Italian, or Hindi, the reader's eye can process the distance between elements and decode them correctly. The top element seems to be a heading or title, while the main element could be text. The three black squares seem like they could be images, since the thin grey lines that are grouped together with them are positioned like captions. The elements could be interpreted differently, but we understand the connections and patterns because the page has been effectively balanced.

In Figure 2.4, we can see that the page or screen has not been used effectively. Items are not equally spaced across the page, and white space does not give a clear sense of which elements are connected to each other. The element in the middle bottom, for example, may be connected to the items on its right, but it may not. Unlike the previous more effective example, this page does not give a sense of what is and what is not connected.

The use of bold, the placement of graphics and images, and the careful use of white (or negative) space are all important ways to account for balance. It is hard to make meaningful sense of the elements on a page or screen if your eye does not travel meaningfully across the various parts. This makes it more likely that your reader will miss key details.

## Alignment

Alignment is a way of showing the organization of the information in your text. Alignment can be either vertical (from the top of the page downwards) or horizontal (across the page, left to right or vice versa).

*Figure 2.3*  Example of effective balance

*Figure 2.4*  Example of ineffective balance

Vertical alignment gives us a sense of where to begin and where to end with a page. It also guides us through the text and helps us to understand how elements are connected and subordinated to each other. A heading is top-aligned, while footnotes are bottom-aligned. We understand the function of both headings and footnotes partly by where they are located. Headings will be discussed in more detail later in the chapter.

Horizontal alignment is either left, centre, or right. Left-aligned elements are squarely against the left-hand margin of the page or screen, while right-aligned elements are squarely against the right-hand margin of the page or screen. Centre-aligned elements are placed an equal distance from both margins. We see alignment in things like headings, which tell us which level of a text we are looking at. Elements like headings will be placed intentionally on the text. Lists, which also use horizontal alignment, will be discussed in more detail later in this chapter.

In Figure 2.5, vertical alignment gives us a clear sense of how the parts on the page relate to each other. We see the title of the document and a heading that separates the last three paragraphs from the first.

The page is clearly broken into two main sections, with vertical alignment guiding the reader through the parts of the text. Horizontal alignment is seen at the paragraph level, where content that needs to be read across the page is positioned accordingly.

Look at the arrangement of elements in Figure 2.6. Does your eye know where to begin? Does your eye 'land' comfortably anywhere on the page? The title and the heading are not logically positioned. The arrangement of the elements in the image does not help you understand how they are connected to each other. The first body paragraph uses centre alignment, which create difficult to scan lines of text. The second paragraph uses justified alignment, which creates larger spaces between some words. This is an example of ineffective alignment. This page does not guide the reader meaningfully across the document.

---

LOREM IPSUM

Lorem ipsum dolor sit amet, consectetur adipiscing elit. Nunc ut lectus quis justo tincidunt sagittis. Vestibulum dolor nisi, hendrerit sed massa eu, blandit malesuada odio. Quisque rutrum iaculis mauris non ornare. Duis sed porttitor mi. Fusce non nibh nec erat pretium blandit eget vel nulla. Sed sit amet ornare nibh. Duis blandit massa in imperdiet efficitur. Suspendisse vulputate, felis at mattis dapibus, felis justo aliquet libero, sit amet semper magna urna sit amet lectus.

*Lorem Ipsum*

Sed vehicula a nulla id fringilla. Maecenas mauris sapien, facilisis volutpat ornare sed, porta vitae massa. Proin bibendum diam quis turpis aliquet volutpat. Donec cursus ultricies neque, ut aliquet nulla vehicula at. Ut cursus placerat dictum. Maecenas vel viverra arcu. Aliquam erat volutpat. Phasellus commodo odio et interdum sagittis. Curabitur vel enim ut orci lacinia rutrum et quis dolor.

Proin vehicula commodo pretium. Etiam hendrerit elit id egestas posuere. Nulla a felis ac turpis tincidunt facilisis eu rutrum leo. Donec vehicula porta posuere. In hac habitasse platea dictumst. Fusce in mattis purus. Vivamus vestibulum ex vel fringilla tempus. Aenean rutrum lacinia tellus sit amet eleifend.

---

*Figure 2.5* Example of effective alignment

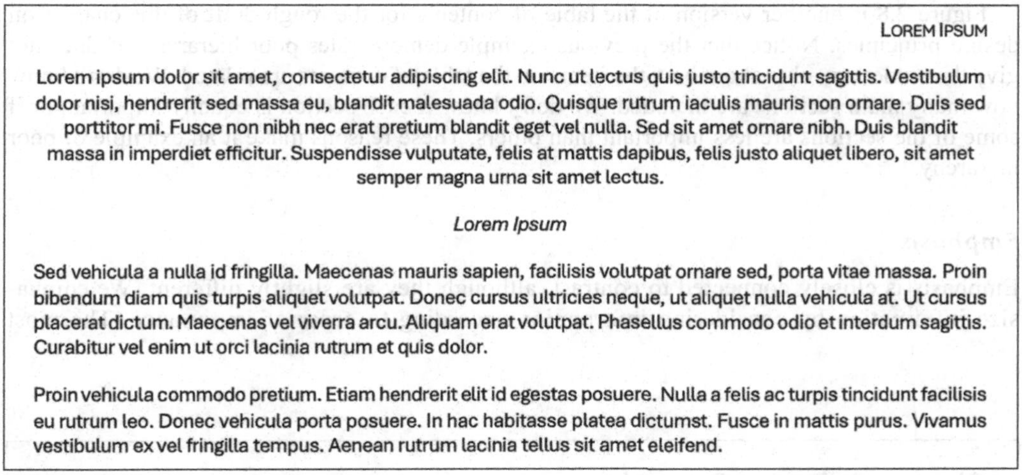

Lorem ipsum dolor sit amet, consectetur adipiscing elit. Nunc ut lectus quis justo tincidunt sagittis. Vestibulum dolor nisi, hendrerit sed massa eu, blandit malesuada odio. Quisque rutrum iaculis mauris non ornare. Duis sed porttitor mi. Fusce non nibh nec erat pretium blandit eget vel nulla. Sed sit amet ornare nibh. Duis blandit massa in imperdiet efficitur. Suspendisse vulputate, felis at mattis dapibus, felis justo aliquet libero, sit amet semper magna urna sit amet lectus.

*Lorem Ipsum*

Sed vehicula a nulla id fringilla. Maecenas mauris sapien, facilisis volutpat ornare sed, porta vitae massa. Proin bibendum diam quis turpis aliquet volutpat. Donec cursus ultricies neque, ut aliquet nulla vehicula at. Ut cursus placerat dictum. Maecenas vel viverra arcu. Aliquam erat volutpat. Phasellus commodo odio et interdum sagittis. Curabitur vel enim ut orci lacinia rutrum et quis dolor.

Proin vehicula commodo pretium. Etiam hendrerit elit id egestas posuere. Nulla a felis ac turpis tincidunt facilisis eu rutrum leo. Donec vehicula porta posuere. In hac habitasse platea dictumst. Fusce in mattis purus. Vivamus vestibulum ex vel fringilla tempus. Aenean rutrum lacinia tellus sit amet eleifend.

*Figure 2.6* Example of ineffective alignment

**Reminder**: alignment is culturally informed. It is affected by the language being used. Languages that are read vertically (like Chinese) will have different alignment expectations than languages that are read horizontally (like English). And languages that are read left to right (like English) will have different alignment expectations from languages that are read from right to left (like Arabic).

## Hierarchy

This is also strongly connected to alignment. Think about the structure of a book: it is often divided into small sections like chapters, and the chapters have further subdivisions. Your reader needs to recognize how these smaller parts relate to the larger parts. A visual hierarchy is a way of representing this. Hierarchy helps the audience to understand which information is superordinate (part of a larger level of organization) and subordinate (contained within the larger level of organization). Hierarchy is about designing a page or screen so that the most important information is seen first. In general, your eye should be guided to the most important element first.

Figure 2.7 is an example using the table of contents for the rough draft of this chapter. It takes into account hierarchy by highlighting the basic structure of the chapter for the reader. It uses a numbered heading system to indicate the five main parts. It uses indentation to show subheadings following this. This shows how hierarchy is connected to alignment. Overall, the improved hierarchy gives the reader a better understanding of how the parts of this chapter work together.

Figure 2.8 is another version of the table of contents for the rough draft of this chapter on design principles. Notice that the previous example demonstrates poor hierarchy. It does not give the reader any clue about how the items on the table of contents are related. We don't know how many main sections are included. We don't know if each section is equally important or if some of the sections are less important than others. These reasons make it an example of poor hierarchy.

## Emphasis

Emphasis is closely connected to contrast, although they are slightly different. We empha-size information by considering information according to order of importance. The most

---

**Chapter 2: Design Principles**

| | |
|---|---|
| Abstract | 1 |
| 2.1  Introduction to Design Principles | 1 |
| Why are Design Principles Important in Technical Communication? | 1 |
| Skimmable Takeaways and Easy Access Points | 1 |
| Chapter Overview | 2 |
| 2.2  Design Principles | 2 |
| Unity | 2 |
| Balance | 3 |
| Alignment | 5 |
| Hierarchy | 7 |
| Emphasis | 8 |
| Proportion | 10 |
| White Space | 11 |
| Repetition | 12 |
| Movement | 14 |
| Contrast | 16 |
| 2.3  Design in IT Contexts: headings | 17 |
| Content-Oriented Headings | 17 |
| 2.4  Design in IT Contexts: lists | 18 |
| Types of Lists | 19 |
| 2.5  Data Visualization Strategies | 20 |
| Entity Relationship Diagrams | 21 |
| Use Case Diagrams | 21 |
| User Flow Diagrams | 22 |

---

*Figure 2.7* Example of effective hierarchy

**Chapter 2: Design Principles**

| | |
|---|---|
| Abstract | 1 |
| Introduction to Design Principles | 1 |
| Why are Design Principles Important in Technical Communication? | 1 |
| Skimmable Takeaways and Easy Access Points | 1 |
| Chapter Overview | 2 |
| Design Principles | 2 |
| Unity | 2 |
| Balance | 3 |
| Alignment | 5 |
| Hierarchy | 7 |
| Emphasis | 8 |
| Proportion | 10 |
| White Space | 11 |
| Repetition | 12 |
| Movement | 14 |
| Contrast | 16 |
| Design in IT Contexts: headings | 17 |
| Content-Oriented Headings | 17 |
| Design in IT Contexts: lists | 18 |
| Types of Lists | 19 |
| Data Visualization Strategies | 20 |
| Entity Relationship Diagrams | 21 |

*Figure 2.8* Example of ineffective hierarchy

important information should draw the reader's eye first, followed by the second most and then the third most. In order for something to stand out, however, something else needs to be less noticeable.

Heading placement and formatting recognizes the importance of emphasis. However, elements like headings can be placed differently on a page and still be recognized according to how important they appear relative to everything else around. With emphasis, everything is relative.

Imagine you are creating a warning sign for the server room door. Which of the following details should be considered most important?

• Dangers caused by frayed wires
• Dangers caused by liquids that are brought into the server room
• Dangers caused by dust
• Dangers caused by overheating

Dangers caused by liquids that are brought into the server room should be emphasized. This is a critical warning because liquids pose a significant risk of short circuits, equipment damage, and potential fire hazards. This is also a preventable hazard. This information should be emphasized on the sign.

On the other hand, while dust accumulation can lead to overheating and equipment inefficiency, it is not as immediate a risk as the hazard posed by liquids. As such, it would not be emphasized as much on the sign.

In Figure 2.9, the hazard of liquids is emphasized by its prominent position in the warning sign and the use of all capital letters to draw the reader's eye to the information. Information has been ordered in a manner that accurately represents the order of importance. The tone is appropriately formal and uses the imperative voice, or command format to clearly indicate that the instruction must be followed.

Example of ineffective emphasis:

Figure 2.10 is poorly designed. It does not effectively emphasize key information. The most important risks are not ordered in order of importance, and strategies to draw attention to them

---

# SERVER ROOM: RESTRICTED ACCESS

# LIQUIDS NOT PERMITTED.

### Please maintain safe operations and report any issues.

---

*Figure 2.9* Example of effective emphasis

---

Server Room Notice

- Please try to keep things tidy.
- Don't bring in drinks if you can help it, but if you do, be careful.
- It gets warm here, so maybe keep an eye on the temperature.
- Sometimes wires wear out, so just watch where you step.
- Dust is not too much of a big deal, but clean if you feel like it.

*Enter at your own risk!*

---

*Figure 2.10* Example of ineffective emphasis

have not been used to highlight these most critical risks. Another problem is that the language is casual in style and tone, which undermines the importance of the warnings, and the most critical risks have not been clearly highlighted.

### Proportion

Remember how emphasis is relative? So is proportion. Proportion considers the elements on a page as they relate to each other. It's about the relationship *between* objects or parts of an object. Proportion is another way of showing hierarchy and alignment. If something is more important, or part of a higher class (superordination), it should be larger than other elements that are part of a smaller class.

Larger text is used to identify headings, while smaller text is used to show the body of a document. Together, they help guide the reader through the document in a logical manner. This is an example of proportion.

Similarly, the size and positioning of images relative to text must be proportionate. When they are carefully sized, relative to text, they complement the text without overwhelming it. If they are too large, they can distract or disrupt the flow of information. If they are too small, they might be overlooked.

Figure 2.11 explains how this paragraph demonstrates good proportion.

In Figure 2.12, we see that the heading is much larger than the text, so it feels disproportionate. As the example paragraph notes, this can contribute to visual confusion for the reader.

### White space

White space is another key design principle that connects to proportion. White space refers to the space you leave empty or blank. It is the negative space that helps us to differentiate between

---

# Heading

This paragraph of text is relative in size and position to the heading. Together, they complement one another effectively. The size of the text is slightly smaller, but it is not so small as to be missed or confusing. It represents what can be considered the Goldilocks of sizing: not too big, not too small – just right.

---

*Figure 2.11* Example of effective proportion

---

# Heading

This paragraph of text is not relative in size and position to the heading. Here, the heading is much, much bigger than the text. This can create visual confusion for the reader. The text may not feel related or connected to the heading. This does not represent the Goldilocks rule. Here, the heading is too big, overwhelming the text, and the text is too small, which can potentially cause it to be left unread.

---

*Figure 2.12* Example of ineffective proportion

parts of a text or screen. It is a key design feature, and it should be used to make a page easier to navigate. More white space between elements makes it faster to navigate around them, while less white space makes it slower. Words are separated by a space and paragraphs are separated by a line space. These are examples of how we use white space to show where one thing ends and another begins.

Proportion and white space work together: the proportion of margins and white space to text impacts readability and how the page looks. Using enough white space helps to reduce visual clutter and makes the document easier and faster to navigate.

In Figure 2.13, white space helps the eye distinguish between parts of the page. The heading and text paragraphs are separated with the negative space, and white space has also been used to border the page.

Figure 2.14, however, represents ineffective use of white space. Not enough negative space has been incorporated into the page. The heading is crowded onto the paragraph, and the paragraphs have not been broken up by a line break. Another problem is the use of justified text, which creates unpredictable spacing between words in the paragraphing. Another problem can be seen in the lack of a margin. Overall, this image feels crowded and visually uncomfortable.

---

**Heading**

Lorem ipsum dolor sit amet, consectetur adipiscing elit. Nulla dictum mi non faucibus eleifend. Class aptent taciti sociosqu ad litora torquent per conubia nostra, per inceptos himenaeos. Nunc in ligula a tortor varius varius in in metus. Maecenas vehicula lobortis congue. Aenean sapien nibh, fringilla eget aliquet id, varius venenatis lectus. Nam viverra eget justo in eleifend. Praesent ligula enim, mollis eget sodales vitae, ornare vitae nisi. Orci varius natoque penatibus et magnis dis parturient montes, nascetur ridiculus mus. Nulla interdum luctus massa eu fringilla. Nam elementum luctus metus, non euismod augue dapibus in.

In a mauris quis nibh porta viverra. Aenean at est et enim consectetur feugiat. Cras ligula urna, tristique eget tellus non, viverra malesuada diam. Nunc accumsan odio sit amet turpis suscipit fermentum. Integer feugiat rutrum dui vitae tincidunt. Phasellus vel mi posuere, maximus libero sit amet, condimentum dui. Etiam tristique tempor neque et porta.

Nunc ex metus, iaculis vitae dictum quis, euismod non diam. Praesent euismod mollis fermentum. Quisque eu euismod turpis. Phasellus mattis velit vitae velit fringilla malesuada. Vestibulum aliquet nunc nec est condimentum molestie. Nullam lorem ligula, tempor fringilla tristique ac, placerat vel ante. Donec molestie magna vel arcu viverra, quis rhoncus lacus congue. Integer vitae lobortis urna, id pretium libero. Integer interdum justo erat, et rhoncus nisl elementum sed. Curabitur mattis ex ullamcorper, rhoncus diam quis, tincidunt felis. Nulla ex ipsum, ultricies non ultrices ac, placerat id elit. Duis ac libero feugiat nisi semper tincidunt. Quisque nec aliquet massa. Integer malesuada congue enim, sit amet semper mi cursus non.

---

*Figure 2.13* Example of effective white space

---

**Heading**

Lorem ipsum dolor sit amet, consectetur adipiscing elit. Nulla dictum mi non faucibus eleifend. Class aptent taciti sociosqu ad litora torquent per conubia nostra, per inceptos himenaeos. Nunc in ligula a tortor varius varius in in metus. Maecenas vehicula lobortis congue. Aenean sapien nibh, fringilla eget aliquet id, varius venenatis lectus. Nam viverra eget justo in eleifend. Praesent ligula enim, mollis eget sodales vitae, ornare vitae nisi. Orci varius natoque penatibus et magnis dis parturient montes, nascetur ridiculus mus. Nulla interdum luctus massa eu fringilla. Nam elementum luctus metus, non euismod augue dapibus in. In a mauris quis nibh porta viverra. Aenean at est et enim consectetur feugiat. Cras ligula urna, tristique eget tellus non, viverra malesuada diam. Nunc accumsan odio sit amet turpis suscipit fermentum. Integer feugiat rutrum dui vitae tincidunt. Phasellus vel mi posuere, maximus libero sit amet, condimentum dui. Etiam tristique tempor neque et porta. Nunc ex metus, iaculis vitae dictum quis, euismod non diam. Praesent euismod mollis fermentum. Quisque eu euismod turpis. Phasellus mattis velit vitae velit fringilla malesuada. Vestibulum aliquet nunc nec est condimentum molestie. Nullam lorem ligula, tempor fringilla tristique ac, placerat vel ante. Donec molestie magna vel arcu viverra, quis rhoncus lacus congue. Integer vitae lobortis urna, id pretium libero. Integer interdum justo erat, et rhoncus nisl elementum sed. Curabitur mattis ex ullamcorper, rhoncus diam quis, tincidunt felis. Nulla ex ipsum, ultricies non ultrices ac, placerat id elit. Duis ac libero feugiat nisi semper tincidunt. Quisque nec aliquet massa. Integer malesuada congue enim, sit amet semper mi cursus non.

*Figure 2.14* Example of ineffective white space

## *Repetition*

Repetition further contributes to understanding how elements are related. It is also important in helping the audience to navigate. We can identify patterns and quickly predict where to find what we need and how subsequent parts will be connected [8]. Repetition is a key to helping achieve this. We see repetition in formatting and placement choices, like page numbers, heading style and emphasis, and even in linguistic features like word patterns (parallel structure). The use of repetition of design elements in a technical document reinforces understanding and reduces cognitive load or the amount of brain power required to read the document.

Figure 2.15 shows how elements like font type and size are used consistently and headings are placed and formatted in the same way to build repetition. We would also expect to see the heading and page number in the same place and in the same style on the following and preceding pages.

Effective repetition can be seen throughout this book. Chapters are formatted in the same manner. The same font choice is used throughout the book, and headings and titles are formatted in the same way. Page numbers are placed in the same place on each page. This kind of repetition reduces the cognitive load on the reader.

When repetition is not used effectively, it creates confusion and unnecessary cognitive processing for the reader. Figure 2.16 uses different font types and sizes. There are also changes to alignment and spacing. Together, these differences create a sense that the document is unfinished or still needs editing. It does not appear to be one cohesive whole.

---

LOREM IPSUM

Lorem ipsum dolor sit amet, consectetur adipiscing elit. In porta, odio ac fringilla maximus, sem mauris iaculis enim, sed pellentesque augue ligula posuere eros. Ut eget ornare eros. Fusce commodo, orci ac vehicula congue, magna lorem vehicula massa, in bibendum quam felis eu est. Vestibulum convallis tempus felis tristique venenatis. In placerat ex urna, suscipit vestibulum massa iaculis eu. Nunc ac velit sed dolor ultricies rutrum. Duis aliquet blandit enim in tristique.

*Heading*
Cras vitae augue volutpat, sollicitudin ligula quis, euismod ipsum. Phasellus id consectetur massa. Suspendisse ex libero, posuere in sem non, congue iaculis leo. Nulla vel ipsum ipsum. Aenean rutrum volutpat augue nec lobortis. Mauris ut tortor efficitur lectus vehicula lacinia sed at dolor. Vivamus porttitor elit ut urna laoreet sollicitudin.

Phasellus suscipit lectus ante, ut scelerisque nulla facilisis at. Vivamus blandit justo varius metus tempor, et dictum erat cursus. Vestibulum semper mi at quam efficitur porttitor. Aliquam ac libero non velit consectetur suscipit at eget diam. Pellentesque sapien enim, euismod eu lobortis condimentum, tincidunt eget ex. Phasellus convallis, ipsum in vestibulum gravida, arcu augue vulputate augue, faucibus porttitor nunc felis et ipsum.

*Heading*
Nam magna nunc, porta ac maximus sit amet, posuere at lectus. Donec vehicula nunc sapien, rhoncus volutpat turpis ultrices id. Duis sollicitudin eget dolor sed facilisis. Vivamus congue metus non feugiat aliquam. Suspendisse porta tempus consectetur. Etiam dignissim nec neque gravida consequat. Proin malesuada, urna eu suscipit feugiat, neque leo pulvinar sapien, id malesuada felis leo non ligula. Donec lorem mi, pellentesque sed venenatis vitae, finibus in arcu.

*Heading*
Morbi a tristique mi. Mauris laoreet nibh in porta gravida. Cras semper odio commodo convallis convallis. Etiam rutrum arcu eu mi venenatis ultricies. Etiam sit amet consectetur dolor, nec blandit dui. Mauris volutpat libero a porttitor porta. Orci varius natoque penatibus et magnis dis parturient montes, nascetur ridiculus mus. Pellentesque id magna in dui commodo faucibus. Ut sollicitudin enim diam, nec placerat nisi commodo id.

*Figure 2.15* Example of effective repetition

## Movement

Remember that alignment and hierarchy are design principles that help to clearly depict how a text or screen is organized. A closely related design principle is movement. Movement refers to how the reader's attention will travel the document. Also closely tied to balance, the eye will naturally land on a specific part of the page [9]. Readers tend to move across a page or screen in a predictable pattern, with some elements attracting more attention. This is culturally informed, however. In languages that write left to right, the eye starts at the top left of the space. In languages like Arabic, the eye starts at the top right. Two patterns that make use of left to write reading patterns are the Z pattern and the F pattern, both discovered in the usability research on eye tracking movements when reading online content [10, 11].

Z patterns are used when a screen or page does not contain a lot of text-based information. This is when the readers eye moves across the top of the screen or page and then scans down the

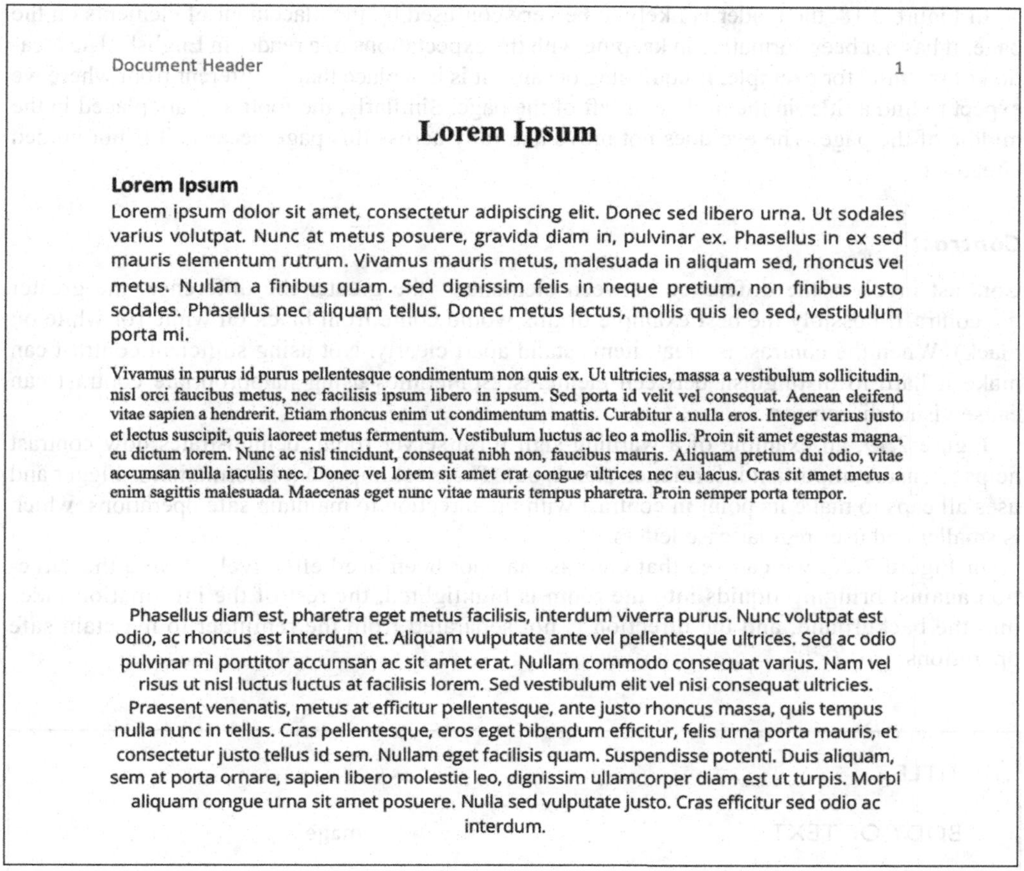

## Lorem Ipsum

**Lorem Ipsum**

Lorem ipsum dolor sit amet, consectetur adipiscing elit. Donec sed libero urna. Ut sodales varius volutpat. Nunc at metus posuere, gravida diam in, pulvinar ex. Phasellus in ex sed mauris elementum rutrum. Vivamus mauris metus, malesuada in aliquam sed, rhoncus vel metus. Nullam a finibus quam. Sed dignissim felis in neque pretium, non finibus justo sodales. Phasellus nec aliquam tellus. Donec metus lectus, mollis quis leo sed, vestibulum porta mi.

Vivamus in purus id purus pellentesque condimentum non quis ex. Ut ultricies, massa a vestibulum sollicitudin, nisl orci faucibus metus, nec facilisis ipsum libero in ipsum. Sed porta id velit vel consequat. Aenean eleifend vitae sapien a hendrerit. Etiam rhoncus enim ut condimentum mattis. Curabitur a nulla eros. Integer varius justo et lectus suscipit, quis laoreet metus fermentum. Vestibulum luctus ac leo a mollis. Proin sit amet egestas magna, eu dictum lorem. Nunc ac nisl tincidunt, consequat nibh non, faucibus mauris. Aliquam pretium dui odio, vitae accumsan nulla iaculis nec. Donec vel lorem sit amet erat congue ultrices nec eu lacus. Cras sit amet metus a enim sagittis malesuada. Maecenas eget nunc vitae mauris tempus pharetra. Proin semper porta tempor.

Phasellus diam est, pharetra eget metus facilisis, interdum viverra purus. Nunc volutpat est odio, ac rhoncus est interdum et. Aliquam vulputate ante vel pellentesque ultrices. Sed ac odio pulvinar mi porttitor accumsan ac sit amet erat. Nullam commodo consequat varius. Nam vel risus ut nisl luctus luctus at facilisis lorem. Sed vestibulum elit vel nisi consequat ultricies. Praesent venenatis, metus at efficitur pellentesque, ante justo rhoncus massa, quis tempus nulla nunc in tellus. Cras pellentesque, eros eget bibendum efficitur, felis urna porta mauris, et consectetur justo tellus id sem. Nullam eget facilisis quam. Suspendisse potenti. Duis aliquam, sem at porta ornare, sapien libero molestie leo, dignissim ullamcorper diam est ut turpis. Morbi aliquam congue urna sit amet posuere. Nulla sed vulputate justo. Cras efficitur sed odio ac interdum.

*Figure 2.16* Example of ineffective repetition

page diagonally before scanning the bottom of the page. A table of contents page is an example of the type of content that would be read by the eye in this manner.

F pattern movement is used when a page or screen has text-heavy content. In this case, the initial movement across the top is then followed by more concentrated horizontal reading across the page at the text-heavy paragraph level. Headings and white space contribute to the F shape seen when this movement is tracked on a page.

Both Z pattern and F pattern movements are effective and should be used for technical documents that are read in English.

Example of effective movement:

Figure 2.17 uses effective movement as elements guide the reader across the page. Remember that images naturally attract the eye, and readers in English naturally start at the top left of the page. Keeping the text on the left ensures that the reader will not miss important information because the eye will naturally start there and then move across the page.

In Figure 2.18, the reader is likely to be very confused by the placement of elements on the page. It has not been formatted in keeping with the expectations of a reader in English. The location of the title, for example, is confusing because it is in a place that is different from where we expect to find a title: in the middle or left of the page. Similarly, the footnotes are placed in the middle of the page. The eye does not move naturally across this page because it is not guided effectively.

### Contrast

Contrast involves the difference between elements – the greater the difference, the greater the contrast. Possibly the best example of this would come from black on white (or white on black). When the contrast is great, items stand apart clearly. Not using sufficient contrast can make it hard to distinguish between elements. Sometimes using inappropriate contrast can cause visual discomfort.

Figure 2.19, an example of a warning sign for a server room, demonstrates how contrast helps to ensure important information stands out. The most important information is bigger and uses all caps to make its point in contrast with the direction to maintain safe operations, which is smaller and uses regular case letters.

In Figure 2.20, we can see that contrast has not been used effectively. While the direction against bringing liquids into the room is highlighted, the rest of the information fades into the background, and the direction is not separated from the reminder to maintain safe operations.

TITLE

BODY OF TEXT                                    Image

BODY OF TEXT                                    Image

footnotes or fine print

*Figure 2.17* Example of effective movement

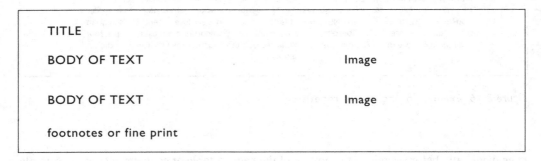

TITLE

BODY OF TEXT                            Image
image
footnotes or fine print
BODY OF TEXT                    Image

*Figure 2.18* Example of ineffective movement

## 2.3   Design in IT contexts: headings

Headings are elements that help readers understand the organization of a text more quickly. Headings are effective when they are mindful of unity, repetition, balance, and contrast. This means they should look like what they are so the reader can use them as stepping stones to skip across the page.

# SERVER ROOM RESTRICTED ACCESS

## LIQUIDS NOT PERMITTED.

### Please maintain safe operations and report any issues immediately.

Figure 2.19 Example of effective contrast

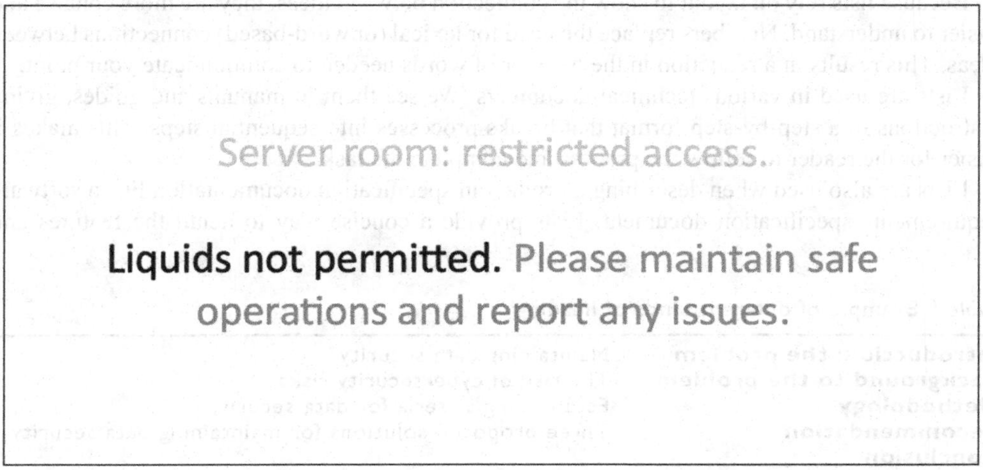

Server room: restricted access.

**Liquids not permitted.** Please maintain safe operations and report any issues.

Figure 2.20 Example of ineffective contrast

Headings help readers navigate across a document, quickly finding the desired information and acting as signposts. Some technical documents are not intended to be read sequentially from start to finish. Headings help readers find relevant sections easily without wasting time with less relevant content. They contribute to the visual hierarchy, as discussed in the design principles, helping the reader to understand how parts of the document relate to each other. This creates a more usable and accessible document.

### Content-oriented headings

Headings in IT documentation should be content-oriented rather than form-oriented [5]. This means that rather than identifying content by its relation to the rest of the text, the heading should identify what information is covered in the relevant section. They should describe and specify the content. However, it is also important that headings are concise. Compare the two examples in Table I.

The example on the left provides a structure-oriented overview of the document. However, it does not provide the reader with a clue about the content of the report. The example on the right, however, gives the reader an idea of what information will be included in the report. This is because it is content-oriented. Wherever possible, it is best practice to use content-oriented rather than structure-oriented headings.

Remember to consider the design principles outlined in the previous section, as they should be implemented to make effective use of headings. Headings should be consistently formatted, contrast from the main content, and demonstrate proportion and emphasis. They should be carefully positioned, incorporating white space and movement in keeping with the expectations of the reader.

## 2.4   Design in IT contexts: lists

Lists are important design features in IT documentation for several reasons. They provide the reader with more information about how to visually process information, how to understand what is prioritized, and clarity about the information being communicated.

Because lists rely on layout to show the connection between ideas, they are more concise and easier to understand. Numbers replace the need for lexical (or word-based) connections between ideas. This results in a reduction in the number of words needed to communicate your point.

Lists are used in various technical documents. We see them in manuals and guides, giving instructions in a step-by-step format that breaks processes into sequential steps. This makes it easier for the reader to follow the process and complete the task.

Lists are also used when describing a product in specification documentation like a software requirements specification document. Lists provide a concise way to detail the features and

*Table I* Example of different kinds of heading

| | |
|---|---|
| **Introduction: the problem** | Maintaining data security |
| **Background to the problem** | The rise of cybersecurity risks |
| **Methodology** | Establishing criteria for data security |
| **Recommendation** | Three proposed solutions for maintaining data security |
| **Conclusion** | |

specifications of a product or service, allowing quick comparison and understanding of its capabilities for users, and evaluation of design for programmers.

Another important use of lists comes in checklists for safety inspections or project milestones. Checklists ensure that all necessary steps or items are accounted for and completed.

Some of the benefits of lists include:

- **Visual organization**: a list lets the reader know how ideas should be communicated in a form that bridges an illustration and text.
- **Reading vertically (scanning)**: a list helps the reader to read down rather than across, which reduces the chances of skipping over items. This makes it less likely that information is missed.
- **Chunking**: when used in instruction manuals, lists provide more digestible chunks of information.
- **Memory recall**: listing information helps to focus attention on key information and the format is easier for the brain to recall.
- **Accessibility**: some readers with cognitive challenges find lists a more accessible means of understanding information.
- **Efficient revision and editing**: lists are easier to edit than information that has been provided as a paragraph. They can be easily modified or updated, which is especially important in technical documents that require frequent updates.

Compare the information in the following non-listed format with its listed format after.

## Non-listed example

Installing a new software application involves several critical steps. Firstly, you need to check the system requirements to ensure compatibility. Once compatibility is confirmed, the next step is to back up existing data to prevent any loss during the installation process. After securing your data, you can proceed with the installation process, which involves running the installer and following the on-screen instructions. During installation, you may be required to select certain preferences and settings based on your specific needs. It is important to keep the system connected to a stable power source throughout this process. Post-installation, it is advisable to update the software to its latest version for optimal performance and security. Finally, you should test the application to ensure it is functioning as expected.

## Listed example

Installing a new software application involves five critical steps:

1  Check the system requirements to ensure compatibility.
2  Back up existing data to prevent any loss during the installation process.
3  Run the installer and following the on-screen instructions.
    Note: during installation, you may be required to select certain preferences and settings based on your specific needs.
    Warning: it is important to keep the system connected to a stable power source throughout this process.

4   Update the software to its latest version for optimal performance and security.
5   Test the application to ensure it is functioning as expected.

Notice the difference in cognitive load and perceived difficulty. The first example is 125 words, while the listed example is only 93 words. No vital information has been removed, but using a list allows the information to be presented in a manner that is easier to understand.

### Types of lists

**Alphabetical lists** provide their items in alphabetical order. These lists are used in indexes, reference lists, and directories where readers are likely to want to find a specific item as quickly as possible and may not be interested in reading any of the other items.

**Bulleted lists** do not require a specific sequence, and they can be read in any order. The hardware requirements and software requirements for a new product or service can be presented using bulleted lists, since they do not need to be read in a specific sequence.

**Chronological lists** provide their items in the order of occurrence. A meeting agenda often presents topics in the order in which they will be discussed. Timelines and lists of historical events are often presented chronologically. In technical communication contexts, a critical incident report would likely include a chronological list of events.

**Nested lists** are used for more complex lists that have subordinate lists below them. Tables of contents in complex documents like this book are a good example. The book is divided into chapters. Chapters are divided into subheadings, and sometimes these are divided into even smaller sections.

**Numbered lists** begin with a number. These are used when content should be ordered in a specific sequence. The steps to follow for installing a new software application should be presented in a numbered format because they must be followed in order.

In summary, lists enhance the readability, organization, and efficiency of technical communication. They make complex information more accessible and easier to understand. Because of this, lists should be used instead of narrative paragraphs wherever possible; however, they should not replace paragraphs that provide an understanding of content or design. In such cases, they should be used to complement the paragraph-based content.

## 2.5   Data visualization strategies

Another key design feature of technical documents is the use of different data visualization strategies. Data visualization is a powerful tool for conveying complex information in a clear, concise, and engaging manner. Some data visualization strategies are common to other forms of professional communication. Graphs and charts can be used to compare, to identify trends and track changes over time. Tables are effective means of presenting text-heavy content in a highly organized format that helps the reader visually understand how data is connected. Infographics combine graphics, charts, and text to present a complex idea in a more readable format. Flowcharts represent processes in a sequential format.

Some data visualization strategies are specific to technical communication documents. Heat maps show how densely data is distributed across different areas or regions. Network diagrams show the interconnections among different components, showing how the different nodes are linked. This is very useful in networking and system design. Decision trees are used to predict outcomes. In data science and machine learning, they are used for classification and regression tasks. They help predict the value of a target variable based on several input variables. Different data visualization strategies are used for different purposes, but together, they form a very important part of technical documents.

### Entity relationship diagrams (ERDs)

In database modelling, one data visualization technique is to create a diagram called an entity relationship diagram (ERD) to illustrate the relationships between different data entities in a system. The visual representation indicates both the element and how it is related. For example, entities are usually rectangular, and relationships could be solid or dashed lines, indicating whether they are mandatory or optional. Knowing this helps the reader to visualize how the parts of the database interact.

**Entities**: objects or concepts about which data is stored. These are usually depicted as rectangles and represent tables in a database. A school database, for example, might include the entities 'Student', 'Teacher', and 'Course'.

**Attributes**: the properties or characteristics of an entity. Usually these are represented as ovals connected to their entity and represent the columns in a table. For example, attributes of the 'Student' entity might include 'Student_ID', 'Name', and 'Date of Birth'.

**Relationships**: how entities interact with each other. They can be visually portrayed as diamonds or lines connecting entities. Relationships can be one-to-one, one-to-many, or many-to-many. 'Student' might be linked to a 'Course' through an 'Enrols' relationship.

**Cardinality and modality**: the nature and degree of relationships between entities. Cardinality indicates the number of instances that one entity can be associated with each instance of another entity. For example, a 'Course' can have multiple students in it, so the relationship would be one-to-many. Modality indicates the necessity of the relationship, showing whether or not an entity instance must participate in the relationship. An example of modality could be 'Mandatory' if the school's policy is that students must be enrolled in at least one course to maintain active status.

**Primary and foreign keys**: primary keys are unique identifiers for each entity and foreign keys are identifiers that enable a dependent relationship between two tables. These are used to show how entities are linked at the database level. 'Student_ID' is an example of a potential primary key for 'Students', while 'Course_ID' is a unique identifier for a specific 'Course'. 'Student_ID' could be a foreign key in a third table outlining 'Enrollment'. Here, it would link back to the table including 'Student' information. Figure 2.21 is a basic entity relationship diagram using these examples.

Entity relationship diagrams are a foundational tool in database design and development. They help developers illustrate the data requirements of a systems and the interrelations of data elements within the database. This is crucial for creating efficient and accurate database structures. The benefit of ERDs is that they help readers to easily understand how the various aspects of a database work together.

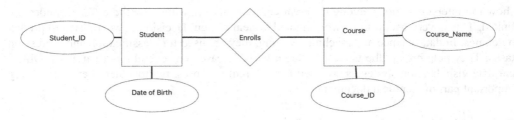

*Figure 2.21* Example of entity relationship diagram

## Use case diagrams

A second type of data visualization strategy involves use case diagrams (UCDs). These are used in software design to help developers understand a system from the perspective of the end user. They show different users and the specific tasks that the user can perform using the application or system.
 Components of a use case diagram include

**Actors**: these are the different users. The word actor is used because it may be a human user or another software system or even an organization. These are usually shown as stick figures.
**Use cases**: these are the specific tasks or functions that the software provides for the actors. They are usually named with action verbs, and they are shown with ellipses in the diagram.
**Relationships**: lines that connect actors to use cases and show how they interact. Use case diagrams include different relationships, such as

- Association – the actor can participate in a use case.
- Extend – one use case extends the behaviour of another use case.
- Include – one use case includes the functionality of another use case.

Figure 2.22 provides a basic illustration of actors within a library management system and the use cases they are connected to.

## User flow diagrams

User flow diagrams are another data visualization strategy used in software development. Sometimes they are also referred to as journey diagrams or UX process diagrams or UX flow charts. They show a step-by-step breakdown of all the actions a user takes when engaging with a product or service. They are helpful in interface (UX) design and considering how easy a product or service is to navigate and use.
 User flow diagrams include the following components:

**Start/end points**: the beginning and ending of the user's actions. They are depicted as circles.
**Steps/actions**: each action the user takes is shown in a process. They are shown as rectangular boxes.
**Decision points**: points where a user might have to make a choice are represented using diamonds. The different choices available demonstrate the possible pathways available.
**Arrows**: arrows connect the different elements and show the direction of the user flow based on the decisions that have been made.

Figure 2.23 shows how IT service request tickets may be processed by an IT help desk support member.

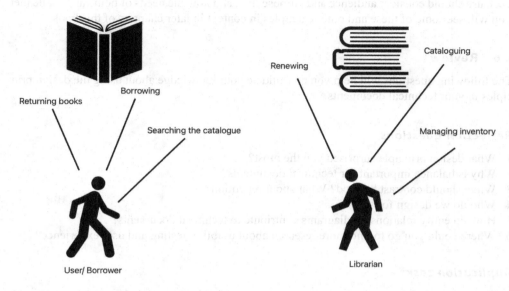

Figure 2.22  Example of use case diagram

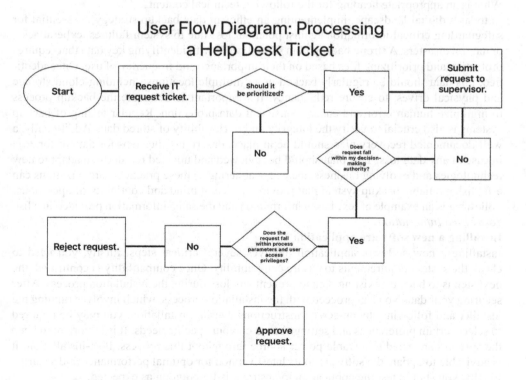

Figure 2.23  Example of user flow diagram

There are many other data visualization tools that are used for different purposes in technical texts. While each have their advantages and disadvantages, it is how they are used that matters. Each use should consider audience and purpose first, and how the needs of both can best be met. You will see some of these and other examples in context in later chapters of this book.

## 2.6   Review

The following questions will help you consolidate your knowledge about using the design principles in your technical documents.

### Reflection questions

1   What design principle surprised you the most?
2   Why is balance important for technical documents?
3   When should contrast be used? What about repetition?
4   Who do we design for?
5   How do entity relationship diagrams contribute to technical documents?
6   Where could you go to find more research about usability testing and user experience?

### Application tasks

1   Create the warning sign for the server room at your school or company. Use appropriate emphasis and other design principles to highlight the most critical hazards as discussed on page 13.
2   What is an appropriate heading for the following technical content?
    In today's digital landscape, implementing an efficient data backup strategy is essential for safeguarding critical information against potential loss due to system failures, cyberattacks, or natural disasters. A strong backup plan should begin with identifying key data that requires protection and prioritizing files based on their importance and frequency of use. Once identified, this data should be regularly backed up to multiple locations, including cloud storage and physical drives, to ensure redundancy. It's important to automate the backup process to minimize human error and ensure consistent data protection. Regular testing of backup systems is also crucial to verify the integrity and retrievability of stored data. Additionally, a well-documented recovery plan should be in place, describing the steps for data restoration in case of any data loss. This plan should be reviewed and updated regularly to adapt to new technologies and evolving business needs. By adhering to these practices, organizations can establish a reliable backup system that provides peace of mind and continuity of operations.
3   Following is an example of text-based information and the same information provided in a list.
    *Text-based Information*:
    **Installing a new software application**
    Installing a new software application involves several critical steps. Firstly, you need to check the system requirements to ensure compatibility. Once compatibility is confirmed, the next step is to back up existing data to prevent any loss during the installation process. After securing your data, you can proceed with the installation process, which involves running the installer and following the on-screen instructions. During installation, you may be required to select certain preferences and settings based on your specific needs. It is important to keep the system connected to a stable power source throughout this process. Post-installation, it is advisable to update the software to its latest version for optimal performance and security. Finally, you should test the application to ensure it is functioning as expected.

*List-based Information*:

Installing a new software application involves five critical steps:

1 Check the system requirements to ensure compatibility.
2 Back up existing data to prevent any loss during the installation process.
3 Run the installer and follow the on-screen instructions.
  Note: during installation, you may be required to select certain preferences and settings based on your specific needs.
  Warning: it is important to keep the system connected to a stable power source throughout this process.
4 Update the software to its latest version for optimal performance and security.
5 Test the application to ensure it is functioning as expected.

**3a** What is the difference between the two in terms of word count?
**3b** What do you notice about repetition of vocabulary?
**3c** What do you notice about sentence length?
**3d** What do you notice about ease of navigation?

4 Review the types of lists on page 20.

**4a** What order has been used to present the types of lists on page 20?
**4b** What is another order that could be used instead? Why?

5 Provide an example of when you would use chronological-based listing in your work or studies.

**5a** Explain why the time-based ordering is important to the context.
**5b** What are some of the consequences of not providing the items in chronological order?

6 Create a better design for the following non-listed technical content. Reformat it, using the design principles and the guidelines for effective headings and lists.
  Configuring a wireless home network requires careful planning and several steps for effective set-up. Initially, you should choose an optimal location for your wireless router, ideally in a central position away from physical obstructions and interference sources like microwaves or cordless phones. The next step involves securing your network by choosing a strong password and selecting the appropriate encryption type, such as WPA3, to ensure network security. After securing the network, connect your devices to the network by selecting the network name and entering the password on each device. It's essential to update the router's firmware to the latest version for enhanced security and performance. Regularly monitoring network activity and changing the password periodically can help maintain security. Lastly, for extended coverage, consider using wi-fi extenders or mesh network systems in areas with weak signal strength.

7 **Review the example ERD about a school.**

**7a** List 3–5 additional entities.
**7b** List several attributes of the entities.
**7c** What are the relationships between the entities?
**7d** What is the cardinality and modality of each relationship?

8 Identify the problems of design with the website wireframe in Figure 2.24.
  **8a** What aspects of unity are problematic?
  **8b** Suggest changes to the website design to improve user experience and align with the design principles.

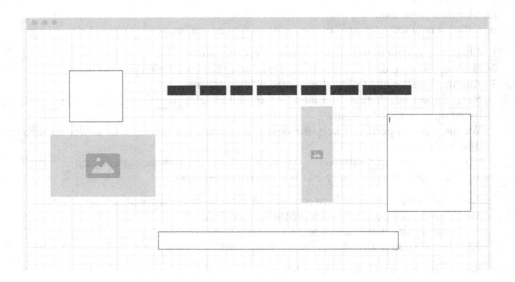

*Figure 2.24* Website wireframe design

9   Create a use case diagram for an automated teller machine (ATM).
   **9a**   Identify all the possible actors.
   **9b**   List the various use cases.
   **9c**   What relationships connect the actors to the various use cases?

## Works cited

1   J. Nielsen, "Usability 101: Introduction to usability," *Nielsen Norman Group*, 03 Jan. 2012. Available: www.nngroup.com/articles/usability-101-introduction-to-usability/

2   B. J. Fogg, C. Soohoo, D. R. Danielson, L. Marable, J. Stanford, and E. R. Tauber, "How do users evaluate the credibility of web sites?: A study with over 2,500 participant," in *DUX '03 Proceedings of the 2003 Conference on Designing for User Experiences*, 2003, pp. 1–15. https://doi.org/10.1145/997078.997097

3   W. D. Ellis, *A source book of Gestalt psychology*. Kegan Paul, Trench, Trubner & Company, 1938.

4   J. Turner and J. Schomberg, "Inclusivity, gestalt principles, and plain language in document design," *In the Library with the Lead Pipe*, 2016. [Online]. Available: www.inthelibrarywiththeleadpipe.org/2016/accessibility/ [Accessed 05 April 2024].

5   J. Balzotti, "Document design," in *Technical communication: A design-centric approach*. Routledge, 2022.

6   D. B. Felker, F. Pickering, V. R. Charrow, V. M. Holland, and J. C. Redish, *Guidelines for document designers*. American Institute for Research, Nov. 1981.

7   L. Itti and C. Koch, "A saliency-based search mechanism for overt and covert shifts of visual attention," *Vision Research*, vol. 40, no. 10–12, pp. 1489–1506, 2000.

8   D. A. Norman, *The psychology of everyday things*. Basic Books, 1988.

9   J. Nielsen and K. Pernice, *Eyetracking web usability*. New Riders, 2010.

10   J. Nielsen, K. Whitenton, and K. Pernice, *The eyetracking evidence: How people read on the web*. Nielsen Norman Group, 2014.

11   S. Djamasbi, M. Siegel, and T. Tullis, "Visual hierarchy and viewing behavior: An eye tracking study," in *Human-Computer Interaction. Design and Development Approaches: 14th International Conference, HCI International 2011*, Orlando, FL, USA, July 9–14, 2011, Proceedings, Part 1 14, pp. 331–340. Springer Berlin Heidelberg.

# Chapter 3

# Workplace communication in IT

## 3.1 Introduction to workplace communication

Whatever your job, you will need to communicate with people both inside and outside of your workplace, be they colleagues, suppliers, customers, or other kinds of stakeholder. To do this well is vital for the success of your job. This chapter will look at the factors which can influence that success, helping you assess the most appropriate choice of communication channel, what the conventions for it are, and the styles of writing you can adopt.

The chapter begins by looking at the different factors to consider when communicating inside your own organization or outside of it. It will then look at a key factor pertinent to *all* communications which is its intercultural aspect. This is followed by in-depth analyses of the audience, purpose, and layout of the most common types of organizational communication – the email, the memo, and enterprise communications. Finally, we look at some of the language features which successful writers use.

### Key vocabulary

*Stakeholder* – a stakeholder is someone who is connected to or involved in a project or business. It refers to people who can affect or be affected by the project or business, as well as those directly working on it.

### Internal vs. external communications

There are different ways to define internal and external communications, and in many companies they are separate branches run by specialists in corporate communications, public relations, and marketing. For such specialists, communications are about producing, conveying, and embodying a brand and developing brand loyalty with a company's various stakeholders, both inside and outside the organization. For some, basing this separation between external and internal audiences on an organization's boundaries is old-fashioned *system* view of communications [1]. It is hierarchical in nature, using memos and newsletters to disseminate an on-brand message, educating the workforce on the objectives of the company, and dividing the company up according to rank and department.

This contrasts with a *resource* model in which the purpose of communications is to promote dialogue and understanding between stakeholders, both internal and external, leveraging that for the benefit of the company. The resource model is less hierarchical, relying on a more organic development of company allegiance, and dividing people according to interests and competencies. Such an approach is fostered by the introduction of enterprise social media or internal

DOI: 10.4324/9781032647524-3

social media. This operates on less of a *push* approach from senior management, and more of a *pull* approach from employees.

What has this got to do with software engineers, though? Well, if you work in IT, these different approaches highlight the choices you need to make about the channels you use and how to communicate to people in those channels. As always, which channel you choose will depend on the purpose and audience of your message. For example, the system approach brings with it a much greater awareness of context and audience, and the kinds of differences in status which often need to be observed when writing to them. This can be useful if expertise needs to be emphasized, such as in the case of a security mandate, when the choice of a memo as a communication tool conveys authority and directiveness.

On the other hand, the resource approach emphasizes what writers and readers have in common. It also helps us understand why some ways of communicating are good for working together as a team and others are good for developing relationships with one person. For example, work chat apps like Slack, Teams, and Yammer are great for finding out who knows a lot about a certain topic. But if you want to ask that person for help, sending an email is a better way to do it.

There are, then, a number of points to take into account when writing internally or externally. Table I outlines some of the key areas:

Overall, then, there is a greater sense of formality and solidity to external communications. However, as we can see from Table I, this often depends on the choice of communication

Table I Factors affected by internal or external communication

|  | Internal | External |
|---|---|---|
| **Audience** | Internal communications are often assumed to be with people-in-the-know, that is colleagues who share your understanding of background context, both for the company generally, and your project. For this reason, they can often be more technical than external communications. | When writing to an external audience, they are less likely to know or understand the internal workings of your company. The opposite is also true in that you may not understand their background, such as their approach to formality. |
| **Purpose** | We often write internally to inform others of what we are doing and have done, or to ask for help of some kind. Providing a rationale is the key to success with both of these because giving reasons for actions makes associated requests more persuasive. | When we write externally, we are usually looking to find a product, a service, or a customer. This means we need to be both polite and clear. |
| **Frequency** | We tend to communicate more often to colleagues inside our companies than to people outside because we are working with them, building projects together. An example of this might be using people's first names. | We write less frequently to customers and suppliers, so we may have less familiar relationships with them. An example of this might be using people's last names and an honorific, or their first names and patronymics. |

*(Continued)*

*Table I (Continued)*

| | Internal | External |
|---|---|---|
| **Style** | Internal communications are often more direct, informal, and use more technical jargon. For example, here the verbs are omitted, and an abbreviation is used: 'any update on the UI?' | External communication needs to promote a good impression of the company, so it tends to be more formal. For example, here the request uses a modal verb (*would*) to make it more polite, and the technical term is said in full: 'would it possible to send an update on the user interface, please?' |
| **Immediacy** | Internal communications range from the synchronous (happening at the same time) to the asynchronous (when reading/listening happen at different times to writing/speaking). Within a workplace, a lot of communication is face-to-face, either in person or online which enables both parties to request and offer immediate verification and elaboration. | External communications can also range from the synchronous to the asynchronous but lean towards the asynchronous because customers and suppliers tend not to exist within the same physical environment. This makes it more difficult for people to ask immediate questions if they do not understand something. |
| **Permanence** | Because internal communications are often face-to-face, they tend not to be part of a permanent record. Even when company meetings are minuted, they do not usually offer a word-by-word account of proceedings. | Given the non-synchronous nature of external communications, what is said is more likely to be written down and therefore be permanently on record. |

channel. Further, in larger companies, there is often a sense that anyone outside of your immediate department is an external customer and is often referred to as such. In this sense, it is wise to treat all internal messages as if they were external messages, writing formally and offering clarity of context at all times.

### Intercultural considerations

Intercultural communication is an important consideration for both internal and external communication, but what is it? These questions and answers should help us understand it more effectively.

**What is culture?** It is a worldview, a set of attitudes, beliefs, and behaviours associated with everything from ethnicity and nationality, through to religion and class which contribute to a person's identity.

**What does it do?** It helps establish a person's sense of what is normal, valued, and appropriate in different situations. For example, when you meet someone for the first time, it is culture which decides whether to bow, rub noses, shake hands, or bump fists.

**So is intercultural communication what happens when people from different cultures interact?** Yes, it is.

**What are the challenges associated with that?** The first challenge is to realize that your way of doing things is not the only way. That is, you need to be alert to the possibility of differences so that you can recognize them. Secondly, you need to respect the differences *as* differences. Bowing and shaking hands are of equal value, and neither are right or wrong, even if one may be more appropriate than the other in different circumstances. The third challenge is to use your cultural sensitivity to anticipate misunderstandings and overcome them.

**That sounds like it might be tricky.** Yes, it can be, particularly as people's sense of who they are and how valued they feel is connected with their culture. Ignoring, disrespecting, or not taking that into account can provoke strong emotions and feel very personal.

**What is the best way to approach this?** By developing your intercultural communication competence (ICC).

**What is ICC?** There is no one set definition, and nor is it just one competence but more like a set of competencies. These include the ability to communicate appropriately and effectively in intercultural situations, recognizing both your own and other world views, and adapting your behaviour accordingly. In other words, someone with ICC considers the areas where miscommunications might happen and acts to prevent them.

**What are the areas where cultural misunderstanding happens?** Probably the most developed version of these areas are Geert Hofstede's dimensions of culture [2]. Hofstede worked at IBM and with other researchers determined the principles underpinning different cultures. These dimensions were expressed along six continuums:

- *Individualism/collectivism* – individualist cultures value independent thought and action, whereas collectivist cultures value interdependent thought and action with an emphasis on what people contribute to their communities.
- *Power distance* – cultures with a low power distance have a more democratic and equitable approach to relationships, whereas high power distance cultures prefer a more hierarchical, uneven distribution of power.
- *Uncertainty avoidance* – cultures with high uncertainty avoidance tend to be more highly regulated with less aptitude for change, whereas low uncertainty avoidance cultures are more comfortable with change and have fewer rules.
- *Long-/short-term orientation* – long-term cultures prioritize the future over the present, and short-term cultures do the opposite.
- *Indulgence/restraint* – this continuum looks at how much a culture allows or does not allow personal gratification of various kinds.
- *Masculinity/femininity* – masculine cultures have more traditional gender roles, valuing ambition and achievement, whereas feminine cultures have more equal ones and value consensus.

All of these dimensions may be expressed more or less explicitly depending on whether the culture is high or low context.

**What are high-context and low-context cultures?** A high-context culture is one where people mean more than they say. In order to understand people in these cultures, you need to consider other contextual clues and non-linguistic pointers, such as tone of voice and body language. A low-context culture is one in which people mean what they say. The way something is said and other contextual factors are much less important [3].

It is helpful to understand that people coming from both types of cultures think they are being clear. It is also useful to remember that people from high-context cultures can feel that people from low-context cultures are acting in a condescending manner to them. Meanwhile, people from low-context cultures might feel frustrated that people from high-context cultures are unnecessarily secretive.

It is also useful to remember that these are not essential descriptions which apply equally to all people in a culture. It is therefore important not to pre-judge someone wherever they come from. Instead, remain alert to the different possibilities on each continuum and try to understand the cultural dimensions at play by listening to the other person.

## 3.2   Email

### Audience

The first point to think about when sending an email is whether it will be urgent for your audience or not. If what you are writing is time-sensitive, then emails are probably not going to be as effective as a phone call because they are less immediate. If your email is not urgent, then it is a good tool to use.

The next point to consider is the context in which the reader of your message will receive it. The great advantage of emails is that they can be accessed, read, and replied to anywhere. Voicemail, videos, and phone conversations, on the other hand, are limited to places where neither volume nor privacy is an issue.

A final point to consider is that email is different from talking in person because when you send an email, you can't clear up misunderstandings right away. Also, because emails are just written words, you can't hear someone's voice or see their face to help understand how they feel or what they mean. This also means emails are less personal than more immediate forms of communication because you cannot change what you are saying in response to the reaction of the person you are speaking with. For some, this results in them feeling that emails are considered less trustworthy than face-to-face conversations.

### Purpose

*The first purpose of an email is courtesy.* By using an asynchronous form of communication, we are saying that the reader can look at its contents when it is convenient for them to do so. This is a recognition that people lead busy lives and either cannot or do not want to respond immediately as they have other pressing matters to attend to. It is common, for example, for employees to restrict the time for reading and replying to emails either to first or last thing each day. You should therefore be prepared to wait for a response for at least 24 hours before following up.

If an issue is time critical, then use the phone, or if it is practical, speak to the person face-to-face. Do not use urgency emojis, such as Outlook's exclamation marks. Firstly, it is rude as you are expressing impatience with someone for something you have chosen to send them. Secondly, it indicates you have not chosen the appropriate medium for your purpose. Email is not an urgent mode of communication. Thirdly, both of these combine to suggest you do not really know what you are doing, tarnishing your professional image.

*The second purpose of email is to provide a written record* of whatever the email is about. One of the disadvantages of face-to-face conversations is that the participants will likely remember what was discussed differently, either in emphasis or detail. It is not possible to relisten to a conversation unless it was taped, which is highly unusual outside of formal interviews. However, when something is written down, all parties can read and re-read it to confirm their understanding or ask for clarification. Emails therefore are a great way to follow up face-to-face meetings, noting the key points for common reference.

*The third main purpose of email is to provide the contents a degree of formality.* Work emails generally use work servers with work addresses. They are part of the official machinery of a company or organization. In this sense, they are explicitly not personal. Before the existence of email, official written communications were conducted using headed paper – paper with the company logo and contact details. Such letterheads gave the letters a look of formality and significance they would not have had without them. Email addresses which use work servers do the same thing. If someone writes to you from a personal email account saying they work for an organization, our instinct is not to trust them. When you write an email, then, you are officially representing the company. As any email may be forwarded to a reader it was not originally meant for at any time, it is worth remembering that.

What are the effects of these three purposes? In terms of content, they mean that emails tend to focus on impersonal matters connected with the business of the organization. In terms of style, or the way they are written, these purposes cause the tone of emails to be more formal, borrowing a lot of the formal staging of letters. Partly as a result of the ability to embed emails in each other as conversational chains [4], there was a tendency in the 2000s-2010s for emails to adopt and adapt some of the informalities of chat messaging services, such as WhatsApp and Telegram. This tendency has decreased with the introduction of enterprise social media.

Email now has a clearer role as a formal means of communication. Given this, the reasons for sending emails tend to relate the following purposes:

- *To inform* (for example, giving new, or confirming previously given, information, like following up after a meeting)
- *To request* (for instance, asking for work to be completed or a meeting to be held)
- *To complain* (such as asking for a response after the expected period of time has elapsed)

### Layout

Given the function of emails as a courteous and formal means of delivering written information, requests and complaints to internal and external stakeholders [5], it is little wonder that they have a well-established layout. This layout consists of the following parts in this order:

1  Subject line
2  Greeting
3  Personalization (optional)
4  Purpose line
5  Context and details
6  Summary (optional)
7  Closing (optional)
8  Signature block

Readers expect most of these parts in an email. If one of them is left out, it will indicate something about your relationship with the reader. For example, if you do not use a greeting, it can mean this is the continuation of an ongoing email chain, or that you are a rude person, or that you know them very well and do not need to be formal with them.

Here is an example of a typical internal email:

Subject: IT alert: scheduled maintenance on Friday[1]

Dear Team,[2]
This is a friendly reminder that we will be performing a scheduled maintenance on our network servers this Friday, 23 June, from 10:00 am to 12:00 pm.[3]

During this time, you may experience some interruptions or delays in accessing the intranet, email, and other online services. Please save your work and log off your devices before the maintenance starts.[4]

We apologize for any inconvenience this may cause and appreciate your patience and cooperation. If you have any questions or concerns, please contact the IT Help Desk at extension 3344 or email ITHelpDesk@suissegator.com.[5]

Thank you for your understanding.
Best regards,

Sara Dais,
Senior IT Services Manager,
IT Services,
Suisse GaTOR Ltd.
+38 455 388392

1  Subject line
A lot of books will advise you to make the subject of your email clear, but what do they mean exactly? What they are referring to is that the purpose of the email should be obvious to the reader. They should be able to understand what you want from them and why you have sent this email. They can then decide what level of urgency and relevance they will assign it. It is valuable to remember that an email is a negotiation not a command. You are in a two-way communication, persuading the reader to give you time, a lot of people's most valuable resource.

So how do you convey effectively and clearly the purpose of your email in the subject line? One way to do this is by using a two-part structure. The first part indicates what type of email it is, e.g., *apology, request, confirmation*, etc. The second part informs the reader of the specifics of the email, e.g., *meeting on Wednesday, Order 2923, conference attendance*, and so on. Follow the first part with a colon, and capitalize the beginning of the second part. Here are some examples:

- Request: emergency leave
- Apology: meeting cancelled on Tuesday
- Complaint: delivery failure of component 334226

2. Greeting
When someone you don't know first writes to you, how do you like to be addressed? In an email the golden rule is never assume familiarity. This is particularly the case when you contact them first. If you are the first person to write the email in an email chain, or it is the first time you have contacted this person, then the level of familiarity has yet to be established. You do not know what level of familiarity the person you are writing to is comfortable with. If it is a professional setting, such as in a workplace or a university, many people prefer to adopt a formal approach to communications, and a more personal and informal tone may not be liked by your reader. Because of this, you should always assume that the person you are writing to wishes to be addressed in an impersonal or formal way. This is particularly the case when you are writing emails for an external readership. With this in mind, begin your emails with *Dear*.

The most formal form of address is to use a title or honorific, such as *Mr.*, *Ms.*, or *Dr.*, and the person's last or first name depending on where you are in the world. If someone has a name you are unfamiliar with, or you are not sure of the gender or gender preference of the person you are writing to, the best approach is just to use their full name – first and last name. This confers upon them a degree of formality while avoiding upsetting them by using the wrong gendered title.

Once the person you are writing to has replied to you, they will let you know how they prefer to be addressed in this set of correspondences. If they write back signing off with just their first name, then you may address them using just that first name. However, if this is an external email and you are writing to a customer or a supplier and it is someone you have yet to meet or establish more familiar ground with, even if they write back using just their first name, you should probably still adopt a more formal approach to them. You are in a professional relationship with them, and it is not an occasion for familiarity.

This is why you should generally use *Dear* rather than other forms of address such as *Hi*. Again, though, you must bear in mind the context. If you work in a very hierarchical organization, for example a government institution of some kind, then addressing your superiors in a more formal fashion will be expected. However, if you work in a company with a flatter structure, then you would be seen as slightly odd if you were using more formal types of address.

3. Personalization

How do you feel when people introduce personal details to business emails? For some cultures, it has traditionally been thought of as offensive to proceed straight to business without asking after someone's well-being first. In other cultures this is considered a matter of form, something that must be acknowledged rather than something anyone cares about. In other cultures, the personal has no place in business and attempts to introduce it are considered rude or even insincere. While cultures differ, so do individuals within that culture.

There are other factors to consider too. For many people there is a big difference between conversing with someone via email or face-to-face, where personalization takes on greater importance. When people are denied the opportunity to meet in person, such as for geographical reasons or during the Covid-19 pandemic, personalization becomes more valued. There are other practical issues. For people who routinely deal with 50+ emails a day, there is often little time for the personal. For more senior colleagues who receive 100s of emails a day, a PA may well answer many of them and therefore personal elements are hard to maintain.

Given these points, it is important to understand the context both when writing and reading emails. As a rule of thumb, it is best to use some kind of personalization when communicating externally. Options include

- I hope this email finds you well.
- I hope your week has got off to a good start.
- I hope you are having a good day.

When writing internally, it is easier to determine the need for personalization. If it is someone you see in person frequently, the less likely it is you will need to add some form of personalization. In both cases, however, if it is the first time you are writing to someone, it is advisable to clarify the context for the reader:

- I was advised by Abdullah AlKetbi from Persec to reach out to you.
- I have been informed by Hwan Choi in Digital Marketing that you are the person best placed to help me.
- I hope you remember me – we met at the IEEE Conference in Hanoi last month.

Sometimes the absence of personalization can be considered a power move, indicating that it is not necessary to be conventionally polite to someone who is junior to them [6].

4. Purpose line

Do you prefer to wait until the final line of an email to find out why someone has written to you, or do you like to know upfront? Most people prefer that emails begin with their purpose. For this reason, the most considerate way to begin an email is by stating why you are writing. You can do this very formally for someone you do not know (e.g., *I am writing to ask if you could please forward me the debugging reports from the last two quarters of 2024*) through to less formally for someone you do (e.g., *Please can you send me the debugging reports from the last two quarters of 2024*). Once you have done this, you can then explain the context and background in the body of the email.

5. Context and details

If you have already stated your purpose in the first line of the email, why do you need to write anything else? This is because people like to know why. The main body of your email can do this, explaining the reasons behind the purpose. Rationale is persuasive, and if you want to persuade your reader, it is important to include it. If you want to call a meeting, inform someone of a service disruption, or ask for information, explaining *why* is likely to make the message more successful. Here are some examples:

- I need the data because I am compiling a coding error summary for Johann Merch.
- This is due to a planned upgrade to the Boss Street servers.
- I think it would be useful to collectively brainstorm key features for Sprint 4, so we can prioritize resources in the spring.

While rationales are persuasive, their effectiveness is reduced by being unnecessarily long. Professional emails are generally short in nature because their purpose is transactional. They are not designed to be novels. If it is vital to go into elaborate detail concerning the content of your email, attach a document. Nobody wants or needs a long email.

6. Summary

When you read an email, particularly if it is a few paragraphs long, do you remember what you're being asked to do and by when? If not, that's what a summary can help with. This is an optional section where you briefly recap the purpose of the email and remind your reader of any deadlines:

- If you could send those figures by 5 October, that would be great.
- Please submit agenda items by 4:00PM on Thursday.
- To summarize, please avoid using the Carson Avenue exit between 08:00–10:00AM on Tuesday 15th March.

7. Closing

Do you like to be thanked when you have done something? Most people do. For that reason, it's courteous to acknowledge the time and effort your reader has gone to, and will go to, both by reading the email but also completing any calls to action you have made. This does not need to be extensive:

- Many thanks for your help.
- Thank you for your time.
- I very much appreciate your help in this matter.

Closing is also an opportunity to demonstrate solidarity with your colleagues. In one example from the research, an English speaker used the closing 'Best wishes and nog veel succes!' to their Dutch colleagues, putting part of the closing in Dutch to demonstrate their collegiality [7, p. 50].

8. Signature block

Have you ever met someone who has the same first name as someone you already know? How about the same first name and family name? Of course, you have. When one of them contacts you via email, how is it clear which of them it is? This is where signature blocks are useful and are even automatically generated in some companies. A signature block is the automated text at the end of an email. With most software you can create a number of options and choose the most pertinent for each email. Ideally, your signature block will include some or all of the following elements:

- Your full name
- Organizational role
- Immediate department
- Company
- Location
- Phone number

An example might look like this:

Dr. Ayo Kehinde,
Director
Research Division
LBB Telecommunications
Lagos
+234 453 988 1218

## 3.3  Memos

A memo (short for *memorandum*) is a brief, written communication usually used within an organization to make announcements, provide instructions, or report recommendations. It is generally not much longer than 1–2 pages and is sent by email as a separate attachment. It is commonly reserved for something more important than daily business.

### Audience

Apart from very specific kinds of memo, such as a Memorandum of Understanding, most memos are designed for internal audiences. This audience is often company-wide, such as when memos are sent out regarding company policy or company vision. However, as many famous examples show, even when memos are designed for internal audiences, they can reach external audiences too. Notoriously, for example, Apple's internal memo asking employees to stop leaking memos was also leaked [8]. This means that although the primary audience is internal, a memo writer needs to remember that anything distributed electronically can easily be forwarded or copied.

Nevertheless, as memos are primarily written for in-house audiences, there can be an assumed understanding about things such as company culture, location, and so on. This does not excuse the need for clearness. Like emails, memos are asynchronous and therefore opportunities for clarification are not immediately available. However, unlike emails they are not usually designed to be responded to. They are a monologue, rather than a dialogue. That is to say they are not meant to be conversational, even if they are conversation starters. Typically, then, they are informative and/or directive.

## Purpose

There are three main purposes for in-house memos: to inform, to direct, and to recommend.

*Informative memos* – these advise employees about new policies or remind them of existing ones, offer company visions, and describe whole-company changes. They are typically written by senior executives if they discuss company culture, or on behalf of departments, such as HR or IT if they explain policies. They generally assume a general audience with no specialist technical knowledge.

*Directive memos* – these ask employees to do something, such as making a password change, or thumbprinting on arriving and leaving. As these are usually connected with company orders, they are often written on behalf of trans-corporation departments, like HR and IT, for general employees.

*Recommendation memos* – these suggest a new idea, approach, or purchase. They are usually more localized than the other kinds of memo and are written for a specific audience at managerial level to aid them in making a decision. While they may include technical elements, the audience is usually non-technical, which is why specialists are the authors, using their expert knowledge to make recommendations for those who do not have it.

## Layout

What does a memo look like? As you can see in the example memorandum, typically a memo looks like a cross between an email and a letter. While content is always more important than style [9], memos are generally split into sections and contain the following elements:

1 Details of record
2 Greeting (optional)
3 Purpose statement
4 Context and rationale
5 Closing (optional)
6 Identifying Information (optional)

## MEMORANDUM[1]

DATE: 17 June 2022
TO: all staff in financial services
CC: Mari Mushref, Chief Executive Officer
FROM: Gentle Pukki, Managing Director of Financial Technology; Ruben Dias, Lead IT Systems Analyst
SUBJECT: data migration Phase 5: legacy shutdown

**Deadline for data retrieval[3]**
With effect from 30 June 2022, the legacy systems will be shut down and decommissioned. Data stored on the older system will no longer be accessible from that point on. It is therefore vital that you complete a final local data audit by that time.

**Rationale[4]**
Following the successful completion of Phase 4 of the Data Migration Project, it is now necessary to shut down the legacy system. This is required in order to maintain data security and reduce costs.

**Follow-up and contact[5]**
If you discover any nonignorable data holes, please contact Ruben Dias, Lead IT Systems Analyst as a matter of urgency on +41 543 7893302 or ruben.dias@swissfintech.com.

1   Details of record

How do you know if you're reading a memo? Unlike other forms of communication, a memo will often announce that it *is* a memo by proclaiming *memorandum* at the top of the page. This does make it easier to find, given how relatively few memos there are, and further, it denotes the formality and importance of the document. Just following that is where the details of the memo are put: who it is to, who it is from, the date, and the subject. The order of these vary according to different templates, but they are the essentials. It is conventional to give people their full name and role in both the *TO* and *FROM* sections, for example:

TO: all employees; Omar Sherzad, C.E.O.; Saif Matsuura, C.F.O.
FROM: Mohammed Al Ketbi, IT Services Director
Similar to emails, a two-part subject line helps make the purpose of the memo clear to the reader *and* the writer, as can be seen in these examples:

- HR Policy Change: maternity leave extension
- IT Security Update: mandatory password change
- Printer Recommendations: feature comparison information for purchase order

2   Greeting (optional)

When communication is not addressed specifically to you but to a group of which you are a part, do you feel annoyed if there is no greeting? This is usually the case with memos. Some memos borrow more features from letters and emails than others and do include a greeting at the beginning. However, this is the exception rather than the rule.

3   Purpose statement

As with emails, it is good practice to state from the very beginning what the purpose of the memo is. This is the reason why someone should read the memo. If you have written a good subject line, you can often just expand that. If we look at the earlier subject lines, here is how they could be expanded for the purpose statement:

- Human Resources is pleased to advise you that maternity leave is being extended from three to six months. This new policy will be effective from 1 September.
- All Costar Group account holders must change their account passwords by the close of business (18:00) today. Failure to complete this will result in immediate loss of access.
- Following the remit established by Aziza Aseel, Chief Administrative Officer, this document establishes criteria for the new printer order, and recommends the most suitable.

4   Context and rationale

Having established what you want to inform your colleagues about or what you want them to do, the body of the memo can be given over to explaining the details. These details can include relevant context but should always provide a rationale – a reason why you are informing, directing, or recommending. The following are examples of such rationales:

- The new policy follows an extensive benchmarking exercise with the other leading international IT services companies to ensure we offer attractive packages commensurate with the world-leading talent at our company.
- The password change is part of a range of measures taken by the IT Security department to ensure that all security protocols are up-to-date across the board.

- The printer purchasing recommendations follows an audit of the company's large-scale printer units which suggests that cost savings can be made by replacing hardware that is largely not fit for purpose, liable to age-related breakdowns, and inefficient.

5  Closing (optional)

If your company sends you information about a new policy, how do you ask follow-up questions? This is particularly important with memos as they are not designed as dialogues, unlike emails. To enable employees to do this, the end of the memo can be used to offer contact details of the person responsible for policy operationalization, or of the website address on the company intranet where clarification can be found.

6  Identifying Information (optional)

Given that the full name and role of the person sending the memo should be at the top of the message, it is unnecessary to repeat it at the end. However, as with letters and emails, it is not uncommon for this to be done, often with additional contact details.

## Recommendation memos

While you may be tasked with writing an informative or directive email, the most likely scenario for IT writers earlier in their careers is to compile recommendation memos. These act as mini reports, often proposing preferred technical solutions to non-specialist managers.

As with other kinds of memos (and indeed reports), the key is to make information clear, relevant, and accessible. You can go a long way to ensuring you achieve these by using a comprehensible and appropriate structure. This might include the following sections:

1  Details of record
2  Greeting (optional)
3  Purpose statement
4  Context and rationale
5  Points of comparison
6  Key conclusions
7  Recommendations

1  Details of record

These are the same as for the other types of memo we have looked at.

2  Purpose

What is the goal of the memo? In this case the objective is to recommend one particular idea, approach, or product above others.

3  Rationale

Why does the company want this recommendation? Briefly describe the background and reasons behind it. It is likely your audience knows why, but as with many other kinds of writing, a great principle is to establish the context clearly so both reader and writer understand each other.

4  Points of comparison

How do you know which idea, approach, or product is the best? Whenever you choose anything, from a movie to a career, you establish criteria or points of comparison and use those to compare which is the best. In this section, then, first list the criteria, and

then compare the leading contenders. This is often most efficiently and effectively done in a table.
5  Key conclusions
   No table explains itself, and it is therefore worthwhile highlighting the most salient points of comparison in a section on its own. What are the extremes for each criterion? Do they focus on one choice?
6  Recommendations
   In the final section, make your recommendation clear and say why.

The example that follows is of a memo seeking to recommend a cloud service host for a mid-sized IT services company.

## MEMORANDUM

DATE: 27 October 2023
TO: Hogan Washman, C.F.O.; Gael Clichy, Director of Strategic Services; Saoirse Murphy, Director of IT Services
FROM: Li Na, Director of Data Division; Jane Rice, IT Liaisons Manager; Chen Long, IT Liaisons Technician
SUBJECT: cloud service host: contract renewal recommendation

### Purpose

The purpose of this memorandum is to recommend a cloud service host.

### Context and rationale

GTT's current cloud service contract with Azul is set to expire in three months. With rapid developments in GTT's product range and the projected expansion into trans-continental markets, it is clear the company is on the threshold of extensive growth. The current contract with Azul is not sufficient to meet expected needs in the medium to long term. Given the difficulties of data migration, it is important to establish the best provider now.

### Points of comparison

The key criteria for choosing a cloud service host are

- Scalability
- Integration
- Uptime and service level agreement
- Business continuity standards
- Business strategy alignment
- Cost

These criteria were benchmarked and referenced against similar size companies in our industry and ranked out of 5.

| | Azul | Nile | Rayo |
|---|---|---|---|
| **Scalability** | 4 | 5 | 4 |
| **Integration** | 5 | 4 | 4 |
| **Uptime and service level agreement** | 4 | 5 | 5 |
| **Business continuity standards** | 4 | 4 | 5 |
| **Business strategy alignment** | 4 | 5 | 4 |
| **Cost** | 4 | 5 | 5 |

*Key conclusions*

All of the top three cloud service hosts score highly across our industry in all categories. Azul is the host which best integrates with our current set-up. However, Nile scores the highest on scalability, and business strategy alignment and Rayo jointly come out top for scores on uptime and service level agreement, business continuity standards, and cost.

*Recommendation*

It is the recommendation of this team that we transfer our cloud service host contract to Nile.

This is because the key factors for us moving forward are likely to be scalability and business strategy alignment, and Nile is the preeminent provider in these categories, as well as being competitive in all other areas.

## 3.4   Enterprise social media

Enterprise social media (ESM) is a term used to refer to in-house social networks. As yet, there is no single established term for ESM as there is for the email and the memo, and other names include enterprise communications, enterprise social networks, employee engagement platforms, employee networking solutions, internal social tools, and more. Some of the most common examples include Slack, Yammer, Chatter, and Jive-n. They range from the more conversational, Twitter-like Slack to the more post-based Facebook-like Yammer. All offer various levels of integration with different services, and others, such as Teams, come with it built in.

*Audience*

The audience for ESM is mostly internal. The more established services can integrate and communicate with each other to enable business-to-business correspondence, but this is the exception. For the large part, then, the audience for in-house social networks can be assumed to have an understanding of context and to be collegial. This enables people on the networks to adopt an informal, conversational tone.

One of the key advantages of ESM is that as well as enabling company-wide communications, it offers departmental and team conversations. This means that when posting in such groups, you can be more certain of the level of technical understanding of your readers. This allows you to use technical terms, common abbreviations, and group references that outsiders would not necessarily understand. This, along with the conversational style of many of the internal social tools means that communication is more efficient and therefore quicker.

With this speed and informality comes expectation of the kind associated with conventional social media, which is that response times should be quicker. While 24 hours may well be a reasonable expectation for an answer to an email, this is not the case for enterprise communications. It is therefore less easy to escape, and responses are usually expected within hours at most.

Another common feature which has arisen from adopting the style of a social network for a work tool is that many people are uncomfortable with the apparent blurring of boundaries between their personal and professional lives [10]. Research suggests that this feeling is most common with younger generations of workers who grew up with social media and view it exclusively as a social concern, only to find that it is now something they have to deal with at work. It is therefore worthwhile remembering that levels of overt sociability may differ from individual to individual and that while ESM usually involves adopting a conversational and personal tone, this is not always welcome.

### Purpose

Research suggests that ESM have been adopted by the large majority of bigger companies, but that it has been done so with no clear aim other than that their competitors have done it. In this sense, ESM has found, and is finding, its own purpose, much as other types of communication have done across the years.

It is a little ironic, then, that one of the key purposes of internal social networks is the accidental discovery. Rather like when we physically meet someone by chance at the coffee machine or in the corridors and chat to them for a few minutes, finding out information about what is happening elsewhere in the company or who is doing what, ESM help the same thing happen virtually. In this way, an internal knowledge network develops from a social network, much as it does outside of work.

Another connected purpose of internal social networks is the distribution of expertise and institutional knowledge. Unlike email, memos, phone calls and other direct forms of communication, internal social tools have an indirect effect too. So-called *lurkers*, those who do not directly engage in a conversation, are able to witness other conversations and learn from them about how to do something or about who in the company knows how to do something [11]. This secret mentoring function enables employees to overcome the borders of their department and their role within it.

A further purpose of ESM is to help people approach other people in a company. Knowing something personal about someone from an internal announcement or their personal page makes it easier to understand if they are approachable [12]. This then helps make a connection with another part of the company, enabling the organization as a whole to develop a greater understanding about itself. This has the effect of minimizing duplication across departments, speeding up resource allocation, and generating greater synergy.

A final and somewhat unhelpful purpose of social networking software is that it becomes a way to estimate value. When someone likes internal social media and contributes a lot to it, those contributions are publicly visible. On the other hand, an employee who does not contribute to it has their lack of contribution shown up by contrast. While enterprise communication contributions are clearly not the measure of an employee's worth, their public visibility can give them a value while leaving quieter colleagues' work somewhat overlooked.

### Layout

Different enterprise communication software works in different ways but most of them enable users to post announcements to different channels and to join in instant message chat streams. These work in the same way as their social networking app counterparts. Two features worth mentioning in this regard are the use of emojis and video messages.

Emojis are commonplace in social chat conversations. They have two main advantages for online communication. Firstly, they enable quick, standardized responses which can broadly express feelings and thoughts more readily than words do. Secondly, emojis compensate for the lack of gestures, such as facial expressions and body language, which we would normally use in face-to-face conversations, and which indicate our attitude towards something. This can be useful to convey intentions, such as humour and surprise. Their use in internal social networks is not as extensive as in external social networks, but they are still able to fulfil these same two functions effectively.

Video messages, made with software such as Loom and CloudApp, or voice notes, made with native applications, can be embedded in chat streams or as separate posts. They are useful ways to guide colleagues through documentation and workflows. They are also a good way to offer a more personal form of message in which you can convey your message with body language and/or say much more in less time if you have an extended message. As with external social network voice notes and videos, brevity is the key to success. It also needs to be remembered that it is not so easy to review something not written down. This means that the purpose of the message should be made obvious at the beginning, and that it should be carefully pronounced, as well as repeated in other words.

## 3.5 Language focus in workplace communications

As we have seen already with our look at intercultural communications, it is vital for effective interaction that we consider the point of view of the person we are writing to. Two ways we can do that is by thinking about *you*.

### Using *You* language

The first way is by using a reader-centred rather than writer-centred focus or what is sometimes called *you* language. This involves looking at things from the point of view of your audience, phrasing things in terms of their interests. Take a look at these contrasting examples in Table II:

In each case, the focus changes from *I* in the initial sentence to *you* in the second, reader-focused version.

Beyond simply changing *I* to *you*, reader-centred language aims to focus on the benefits of an action or request for the reader, rather than the writer. Look at the following ways of making the same request:

1 I want you to finish that report by Friday.
2 The management team would appreciate your efforts if you could get that report off your to-do list by Friday.

Here a) is a straightforward demand, whereas b) not only underlines that the report will be useful to the management team, thereby giving a reason why it needs to be completed, but also

*Table II I* vs *You* language

| *I language (writer-centred)* | *You language (reader-centred)* |
| --- | --- |
| I apologize for my late response | I hope the lateness of my reply has not inconvenienced you unduly |
| I have attached the file | As requested, please find the file attached |
| If possible, I'd like to know more about . . . | It would be a great help if you were able to explain more about . . . |

suggests it would relieve the pressure on the report writer too. Here is another example of saying the same thing in two different ways:

1  I am sorry that I will not be able to finish the report by Friday.
2  I apologize for not being able to complete the report by Friday, but I hope you can understand that it is better to have a fully fact-checked report on Monday than one which may be compromised one working day earlier.

Here a) is an honest apology, but b) gives the reader a reason to accept the apology, a reason that is in the reader's best interests.

### Not using You language

Unlike many languages, English does not have a polite form of *you*. The pronoun *you* can be used for one person or many people, and it can be formal or informal. This ambiguity means it is sometimes open to being misread by readers as a singular, personal and informal *you*, even if that is not what is meant. In such cases it can sound like the writer is accusing the reader of being at fault. One way around that is to remove personal pronouns altogether, or replace them with impersonal entities, such as departments or companies, and to use the passive voice. Consider the following statements in Table III:

*Table III* Not using *You* language

| Faulty You | Impersonal and/or passive |
| --- | --- |
| Until you complete the coding, we cannot proceed. | Until the coding is complete, the project cannot move forward. |
| We are concerned about your delivery schedule. | Compu Inc. is concerned about Cloud Serve's delivery schedule. |
| I need to check the updated projections you gave me. | The updated projections need to be checked by the Security department. |

In each case here, the *you* element is removed whether by using the passive voice (to be + past participle), by replacing it with proper nouns (names of companies or departments), or by doing both. This removes the possibility of assigning blame in a personal way.

### Avoid negative phrasing

A final area where we can improve relationships across the written divide is by avoiding negative phrasing. This can involve using positive words rather than negative ones, but also focusing on solutions rather than problems and maintaining an openness to feedback. Have a look at these examples in Table IV:

*Table IV* Avoiding negative phrasing

| Negative phrasing/focus | Positive phrasing/focus |
| --- | --- |
| This is impossible. | Let's work together to find a solution. |
| You're doing it wrong. | How is that being done? There is probably a way to solve it. |
| You need to follow my instructions. | The process I sent is yet to work as it should. Can you let me know where the issue might be? |

In each of these examples, there is an attempt to reformulate the phrase in a more positive manner which focuses on the issue and invites collaboration.

## 3.6    Review

The following questions will help you refresh your knowledge about workplace communications and deepen your understanding through practice.

### Reflection questions

1  What are the key differences between internal and external workplace communications?
2  What are the challenges when communicating across cultures?
3  How can those intercultural challenges be overcome?
4  When is *not* communicating via email appropriate?
5  What are the three main purposes of email?
6  Who is the audience for most memos?
7  What are the three main types of memos?
8  How does ESM help the spread of expertise inside a company?
9  What are the advantages of emojis in communication?
10  In what ways can we consider the point of view of the reader when we write?

### Application tasks

**1a**  Profile your cultural dimensions in Table V, giving examples of when you have communicated in that way.

*Table V* Personal intercultural profile

| Dimension | Answer | Example |
| --- | --- | --- |
| Are you more of an individualist or a collectivist? | | |
| Do you prefer a low power distance or a high power? | | |
| Are you comfortable with uncertainty or more comfortable avoiding it? | | |
| Do you have a long-term or short-term orientation? | | |
| Do you prefer indulgence or restraint in your work life? | | |
| Do you have a more masculine or feminine approach? | | |
| Do you communicate in a high-context or low-context manner? | | |

**1b**  Think of three challenges your approach to communication might cause for someone of the opposite profile. Then write down in Table VI how you would deal with each challenge to make your communication more successful.

Table VI  Personal intercultural challenges and solutions

| Challenge | Solution |
|---|---|
| 1 | |
| 2 | |
| 3 | |

**2a** You need to suspend HR e-services for the company you work for while you address a security breach. Consider the following factors and make notes in Table VII.

Table VII  E-services suspension communication factors

| | Questions | Answers |
|---|---|---|
| **Audience** | Do the people you're writing to share the same background context as you? Is the audience technically knowledgeable? Is the audience senior to you? | |
| **Purpose** | Are you writing to inform, make a request, or apologize? Do you have a rationale? | |
| **Frequency** | Is this someone you write to frequently? Will they know who you are? | |
| **Style** | Do you need to write in a formal or informal tone? | |
| **Immediacy** | How time-sensitive is your communication? Will the audience be able to ask for immediate clarification? Will you be able to offer immediate elaboration? | |
| **Permanence** | Does your communication need to be retrievable? Does your communication need to be editable? | |

**2b** Given your previous answers, rank the following communication channels in Table VIII, with 1 = the most appropriate, and 6 the least appropriate. Give a reason for each.

Table VIII  E-services suspension communication channel choice

| Channel | Rank | Reason |
|---|---|---|
| Email | | |
| Internal social media | | |
| Letter | | |
| Memorandum | | |
| Phone call | | |
| Video message | | |

**3a** You need to meet with the colleague who previously held your position to help you understand one of the processes involved in a data migration project. You have never met her before. She has been promoted and moved to your company's offices in Dubai. You decide to write her an email. What would be a good subject line? Why?

*Subject line*
1 Meeting request
2 Meeting request: data migration project
3 Can we discuss the data migration project?

**3b** You now need to compose the rest of your email. Make the appropriate choice for each part.

*Greeting*
1 Dear Sophia
2 Dear Mrs. Markovka
3 Dear Ms. Markovka

*Personalization*
1 I hope this email finds you well.
2 I hope Dubai is treating you well.
3 –

*Purpose line*
1 I am writing to see if we can meet this week to discuss the data migration project.
2 I am working on the data migration project.
3 Can we have a meeting?

*Context and details*
1 I took over from you when you moved to the Dubai offices. I am a bit stuck.
2 I am now performing the role on the project you had before you abandoned us. I am unsure of certain of the procedures, and I need help.
3 You did a great job when you were in the role I'm now in, and I was wondering if you could share some of your expertise. I understand you are particularly knowledgeable about Phase 3 of the project.

*Summary*
1 –
2 If you are free any time before midday Dubai time this week, let me know which day suits you best and I will set up a Zoom call.
3 Tuesday afternoon from 3–5 is best for me, so hopefully you can make it then.

*Closing*
1 I can't wait to hear about the shopping in Dubai!
2 –
3. Many thanks for your help.

*Signature block*
1 Yours sincerely,
  Mima Ito

2 Kind regards,
  Mima,
  Lead Technician,
  IPPE,
  Tokyo
3 Best regards,
  Mima Ito,
  Lead Technician,
  IPPE,
  Tokyo

**3c** Before you send the email, a colleague tells you that Sophia is a low-context, low power distance, individualist. Would you change anything about your choices?

## Works cited

1  R. Chen, "The stakeholder-communication continuum: An alternate approach to internal and external communications," *Journal of Professional Communication*, vol. 6, no. 1, pp. 7–33, 2020. https://doi.org/10.15173/jpc.v6i1.4350
2  G. Hofstede, "Dimensionalizing cultures: The Hofstede model in context," *Online Readings in Psychology and Culture*, vol. 2, no. 1, pp. 3–25, 2011. https://doi.org/10.9707/2307-0919.1014
3  E. T. Hall, "Unstated features of the cultural context of learning," *The Educational Forum*, vol. 54, no. 1, pp. 21–34, 1990. https://doi.org/10.1080/00131728909335514
4  J. Gimenez, "Embedded business emails: Meeting new demands in international business communication," *English for Specific Purposes*, vol. 25, no. 2, pp. 154–172, 2006. https://doi.org/10.1016/j.esp.2005.04.005
5  S. Park, J. Jeon, and E. Shim, "Exploring request emails in English for business purposes: A move analysis," *English for Specific Purposes*, vol. 63, pp. 137–150, 2021. https://doi.org/10.1016/j.esp.2021.03.006
6  P. Millot, "Inclusivity and exclusivity in English as a Business Lingua Franca: The expression of a professional voice in email communication," *English for Specific Purposes*, vol. 46, pp. 59–71, 2017. https://doi.org/10.1016/j.esp.2016.12.001
7  C. Nickerson, "The use of English in electronic mail in a multinational corporation," in *Writing business: Genres, media and discourses*. Routledge, 2014, pp. 35–56.
8  M. Gurman, "Apple warns employees to stop leaking information to media," 13 April 2018. [Online]. Available: www.bloomberg.com/news/articles/2018-04-13/apple-warns-employees-to-stop-leaking-information-to-media?embedded-checkout=true [Accessed 25 March 2024].
9  N. Amare and C. Brammer, "Perceptions of memo quality: A case study of engineering practitioners, professors, and students," *Journal of Technical Writing and Communication*, vol. 35, no. 2, pp. 179–190, 2005. https://doi.org/10.2190/ML5N-EYG1-T3F7-RER6
10 H. Koch, D. E. Leidner, and E. S. Gonzalez, "Digitally enabling social networks: Resolving IT – culture conflict," *Info Systems Journal*, vol. 23, pp. 501–523, 2013. https://doi.org/10.1111/isj.12020
11 P. M. Leonardi and E. Vaast, "Social media and their affordances for organizing: A review and agenda for research," *Academy of Management Annals*, vol. 11, no. 1, pp. 150–188, 2017. https://doi.org/10.5465/annals.2015.0144
12 N. Boukef, M. H. Charki, and M. Cheikh-Ammar, "Bridging the gap between work- and nonwork-related knowledge contributions on enterprise social media: The role of the employee – employer relationship," *Information Systems Journal*, pp. 1–41, 2024. https://doi.org/10.1111/isj.12500

# Chapter 4

# Process documents in SDLC

## 4.1 Introduction to project management documentation

This section looks at the kinds of documents that are used throughout the software development cycle (SDLC). The SDLC describes the development of a project or even a feature from the initial brainstorm and planning through coding, testing, release, and maintenance.

### Types of project management

A cycle is a regularly repeated set of events. The cycles in SDLCs can be short, such as a couple of weeks or months or much longer. These differences in cycle length are associated with different approaches to project development. Since the early 2000s, many software projects have been structured around *agile* development. Agile development is often contrasted with *waterfall* development. Although neither of them is a single entity and they are not always completed in the same way, we can identify some core features of each in Table I.

Different approaches are suitable for different types of projects. If the client has a very clear idea of how the project should proceed or safety and legislation are vital, such as in the aerospace industry, then the waterfall approach is more appropriate. If the client needs something quickly to test the market and is happy with an incremental approach to feature delivery, then the more flexible agile approach is a better fit.

What this means for documentation is that while no set of documentation is ever complete, in the agile approach it is subject to constant revision, whereas waterfall approaches develop more permanent documentation.

In both cases, although they appear to be driven by coding, SDLC documents play a crucial role in successful cycles.

### The value of SDLC documents

As an engineer, you may feel that writing documents is not your job and that the code is the only documentation needed [1]. However, there are many reasons why creating and keeping good SDLC documentation is beneficial to you, to the project, to your clients, and to anyone who wants to understand what you did:

- **Guides project vision.** Good documentation for the SDLC helps with the development process and also guides it [2]. Usually, the plan for the project is written down in a document before anyone starts writing the actual programme.

DOI: 10.4324/9781032647524-4

*Table I* Waterfall vs. agile development

|  | *Waterfall development* | *Agile development* |
| --- | --- | --- |
| **Time** | Longer cycles for bigger projects | Based on sprints – short cycles of two weeks – two months |
| **Flow** | One direction | Iterative, circling back again and again |
| **Dependencies** | Each step follows the previous one and cannot happen without it | Steps, like development and testing happen simultaneously |
| **Deliverables** | The project is completed at the end | The project delivers parts of the product on a regular cycle |
| **Client input** | At the beginning | Continuous |
| **Budget** | Fixed | Flexible |
| **Documentation** | Established at the beginning and end of the project | Continuously revised and minimized |

- **Establishes project agreement.** Good documentation ensures that the vision of the different stakeholders is the same as the developers [3]. It becomes a reference point against which conflicts of understanding can be addressed and agreement reached.
- **Helps collaboration.** SDLC documentation enables better collaboration between developers. For example, other engineers are able to see the approach to a problem and offer more efficient or cost-effective ways to solve it.
- **Records project history.** As SDLC documentation is usually a shared effort and available for all stakeholders to see, it becomes a way to protect yourself too. People can check what you are doing and have their opportunity to contribute, and if they do not, you are able to point to that in the document.
- **Smooths developer onboarding.** SDLC documents allow developers joining your project to get up-to-date on everything that has been done and is being done more quickly.
- **Enables project referencing.** SDLC documents are a way for people to reference your work. They can see what you are doing and what you have done, and it enables them to talk about it with you and with others more easily.
- **Showcases your work.** SDLC documents can be a showcase for your contribution, demonstrating to others your impact on the project.
- **Assists future development.** Having good documentation also extends the lifetime of a project [4]. It enables you, other developers, or stakeholders to return to a project long after it has finished in development. Everyone can see what the objectives were, how they were achieved, and why. This legacy action is far more effective for future developers than trying to work everything out by deduction.

The common feature of all of these advantages is that they save you time. Instead of having to explain what and why and how and when you did things, over and over again, these are all in the documentation. Good documentation is therefore invaluable.

## Different types of SDLC document

SDLC documents are sometimes separated into product and process documents:

- **Process documents** are to establish, maintain, and record the process of product development. They include planning documents, roadmaps, and coding standards. We are going to look at process documents in this chapter.

- **Product documents** are used to describe the product being developed. These documents are split between those designed for use by developers, such as requirement documents (see Chapter 5), and those created for users and system administrators, such as instruction manuals (see Chapter 6).

In agile development processes, all of these documents are likely to be written and rewritten multiple times, but in waterfall development they are designed to be written once and then subject only to minor updates.

## 4.2   Work breakdown structures

This section looks at work breakdown structures, who uses them, why, and how.

### What is a work breakdown structure?

A work breakdown structure (WBS) is a way of breaking a larger project into smaller tasks [5]. For example, let us imagine a very simple project, like making a cup of tea. This can be broken down into a series of smaller tasks:

1  *Making a cup of tea*
   a  Boil water
   b  Put tea bag in cup
   c  Pour water into cup
   d  Wait two minutes
   e  Remove tea bag

As you can see, the WBS for making a cup of tea starts with the objective and then moves backwards to list all the steps required to reach it. Each of these steps could be broken down further. For example, what do we need to do to boil water?

1  *Making a cup of tea*
   a  Boil water
      i    Turn on tap
      ii   Run water for 30 seconds
      iii  Fill kettle
      iv   Put kettle on stove
      v    Turn on stove

Breaking down a project aim like this is helpful in a number of ways:

- **It allows the project team to determine the scope of a project.** For example, do we have a tap for the water? Do we need to buy one or build one?
- **It shows the steps required to reach an objective.** Boiling water is one step by itself and does not look challenging, but it is composed of five steps in turn allowing us to understand that it requires much more time and effort.
- **It helps us understand what resources are needed and where.** Once we have broken all the steps down into subtasks, we can see that boiling the water is the most resource intensive part of the whole tea making project. We therefore need to give more resources to that part.

- **It focuses on deliverables, from the smallest task all the way up to the aim of the whole project.** Each task and sub-task are about completing an action, such as filling the kettle. It therefore makes it easier to understand what jobs need doing because at its simplest, a WBS is a to-do list.
- **It enables us to assign tasks fairly and appropriately.** Having a list of tasks lets us see which ones are more demanding and which require particular skill sets. For example, if we need someone to turn on a stove, that person needs to have experience with cookers, possibly gas and electric. Do we have a person like that on the team?

### Who uses a WBS?

It is normally a project manager who uses a WBS. However, it is important to understand that most companies do not employ specialist project managers. Rather, it is a role that employees from other specializations take on.

A WBS is normally made at an early stage in an SDLC. Waterfall projects will likely try to do this once at the start of a project, whereas agile projects need to revisit this for every new sprint cycle. Regardless of the type of project management, a WBS relies on extensive consultation with all the people involved in the project. Only a specialist will know, for example, whether existing library code can complete 40% or 80% of a specific task.

Once a WBS has been created, it is communicated to a project team in a number of different ways. We will look at two of them here: *Gantt charts* and *Kanban boards*.

### Gantt charts

A Gantt chart is a form of table which shows a WBS. Over the years, it has remained as one of the most widely-used documents in IT project management [6–8]. It is a forward-looking document outlining expected tasks, completion schedules, and task ownership. Table II is an example of a simple Gantt chart.

**WBS #** – this stands for Work Breakdown Structure Number. Each task is given a number for reference and to indicate its place in the hierarchy of the project, whether that be in terms of value or more usually chronology.

**Task** – the task indicates the part of the project. These parts have different names:

- *Summary tasks* – these are top-level tasks or headings. In Table II the summary task is Help-desk Software Design, Development, Deployment.
- *Subtasks* – these are the level below summary tasks. In Table II all the tasks with decimal points (5.1–5.9) are subtasks.
- *Work tasks* – these are the level below which there are no more levels and describe the actual work that needs to be done. They do not feature in Table II because it is not that detailed. On most Gantt chart software, work tasks are only visible after clicking on chevrons or links to access that level. This is because a Gantt chart generally gives a broad overview of a schedule. In a sense, Gantt charts are ways of representing or viewing a WBS. They look at a WBS from an overhead view, giving the audience the big picture rather than a detailed one.

**Task owners** – this column indicates who is responsible for each task. For instance, in Table II the design is the responsibility of the developers and the quality assurance engineers. In this way, everyone on a team knows what they have to do, and the rest of the team knows who they need to ask about that particular part.

*Table II* Part of Gantt chart for software development project

| WBS # | Task | Task owners | Start date | End date | Aug | Sep | Oct | Nov | Dec |
|---|---|---|---|---|---|---|---|---|---|
| 5 | **Design, Develop, Deploy Helpdesk Software** | All Teams | 2024-08-01 | 2024-12-30 | | | | | |
| 5.1 | **Gather requirements** | Product Owner, Devs | 2024-08-01 | 2024-08-14 | | | | | |
| 5.2 | **Design** | Devs, QA Engs | 2024-08-05 | 2024-09-10 | | | | | |
| 5.3 | **Develop** | Devs, QA Engs | 2024-08-01 | 2024-10-31 | | | | | |
| 5.4 | **Document** | Doc Team | 2024-10-10 | 2024-11-15 | | | | | |
| 5.5 | **Train** | Training Team | 2024-11-05 | 2024-11-30 | | | | | |
| 5.6 | **Test** | QA Engs | 2024-10-01 | 2024-11-20 | | | | | |
| 5.7 | **Deploy** | Deploy Team | 2024-11-21 | 2024-12-24 | | | | | |
| 5.8 | **Monitor** | Monitor Team | 2024-11-21 | 2024-12-30 | | | | | |
| 5.9 | **Support** | Support Team | 2024-11-21 | 2024-12-30 | | | | | |

**Start/End dates** – this is the scheduled calendar period for the task to be completed. This does not mean that the task takes this long, but that this is the period allotted for it. Very often, people work on more than one project at once, or they will have routine tasks that mean they cannot devote all their energies to a project. Usually, calendar time is a lot longer than the time required to complete the task.

**Months** – Gantt charts usually split time into manageable portions, ranging through weeks, months, and quarters (three months). What this part of the chart reveals is the linear progress of a project, and quite often the dependencies too. A dependency is a task that needs another task to be finished before it can be done. For example, in Table II *Training* cannot begin until *Development* is complete. However, Gantt charts only convey a sense of dependencies. On time-critical projects, a dependency or network diagram conveys this more accurately. It reveals potential problems, as well as the *critical path* or shortest possible route where the start of one part depends on the finish of another.

In summary, a Gantt chart is a graphic representation of a project plan. It is a cross between a to-do list and a calendar. It usually involves some or all of the following communicative features:

- **Verbs define tasks.** A noun is what something is, but the task is a verb. For example, an API is a noun, but not a task, whereas *developing an API* is something that can be done. Verbs take time, whereas nouns do not. In Table II, for example, 5.1 is *Gather requirements*, not *Requirement gathering.*
- **Abbreviations keep charts short.** You can use abbreviations for anything that is commonly understood, such as months in Table II. This helps to keep Gantt charts more efficient and

allows them to achieve one of their key aims which is to provide stakeholders with an overview of a project.

- **Time moves horizontally.** This is a concept we are used to. We think of time moving forward, not up and down. Different cultures will represent this as a left or right movement.
- **Tasks countdown vertically.** The first task is at the top of a table, the second underneath it, and so on. This gives many Gantt charts the so-called waterfall look, as they cascade on a slope, left to right in this case. In agile projects, however, there are fewer dependencies and tasks can happen simultaneously, giving the chart a more tower block appearance. Looking at the profile of tasks on a Gantt chart should therefore help you understand the kind of project management being used.

## Kanban boards

Although Kanban boards were originally created for the car industry, in recent years they have been adopted by IT departments and companies. They are a way of visualizing workflow [9]. Unlike Gantt charts, which are more forward-looking, Kanban boards are more focused on what is happening now and what is waiting to happen. This makes them a great way to map the current situation graphically.

*Table III* Sample Kanban board

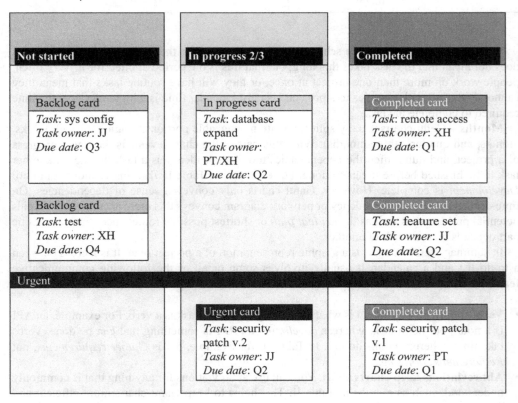

**Columns** – Kanban boards are split by columns. The most basic boards, such as the one in Table III, are divided into three parts.

- The first column from the left, here called *Not Started*, shows those tasks which are back-logged or which have been accepted but not yet started.
- The second column, here called *In Progress*, lists those tasks which are currently being worked on.
- The third column, called *Completed* here, outlines the tasks which have been completed.

Together, these columns show the workflow of a unit or department at this current point in time.

**Work-in-Progress Limit** – at the top of the *In Progress* column is the WiP Limit. This iden-tifies the maximum number of tasks that can be worked on at any one time. In Table III the limit is three tasks, and two are being worked on currently, hence 2/3.

**Kanban cards** – the cards in each column show the tasks appropriate to that point in the workflow. For example, there are two cards which have yet to be started in the *Not Started* col-umn. Kanban cards are usually double-sided:

- *The front side* gives you basic information about the task, such as what the task is, the sub-tasks, who the task owner is, and the length of task time.
- On most Kanban software, *the reverse side* usually allows for others to leave comments, files, links, and other connected data which is relevant to the task. This aligns with a Kanban's live status. It is not just a planning board, nor a record of achievement, but rather a means of keeping everything clear so everyone connected to the project can be instantly up-to-date.

Although most Kanban cards are individually configurable, typically, there are templates for both sides of the cards. These are task specific and save time when dealing with repetitive tasks, such as debugging.

**Swimlane** – the swimlane on a Kanban board is a horizontal line used to differentiate teams or kinds of task. In Table III the swimlane is marked *Urgent* and is used to distinguish priority tasks from tasks in the standard workflow. Sometimes Kanban boards are split by swimlanes into three parts, with the topmost being the highest priority level of task, followed by normal and low priorities. This kind of board is common in IT operations where the number of tasks almost always exceeds the time available.

As with Gantt charts, Kanban boards look at time horizontally and tasks vertically. They also use abbreviations but are more likely than Gantt charts to use noun phrases rather than verbs. This is because Kanban boards are more often used by small teams or departments where everyone is a specialist. Noun phrases are more difficult to understand for non-specialists as the action, the verb part, is hidden in the noun.

## 4.3    Roadmaps

Roadmaps are ways of plotting the desired progress of a project. As the name suggests, they show users how to get from the beginning of a project to the end, just as you would with a nor-mal map.

As with normal maps which look at the same terrain in different ways, there are a number of different project roadmaps. Three of the most common kinds in the SDLC are technical, strate-gic, and release roadmaps, together known as product roadmaps.

### Strategic roadmaps

A strategic roadmap is an outline of three main parts of a project:

* The vision
* The aims
* The steps

A strategic roadmap is generally created at the start of an SDLC, but it is open to change during the lifetime of the project.

### Audience

A strategic roadmap is designed for all stakeholders, including clients, senior management, and investors, as well as the project team. It therefore needs to take account of two aspects key to these different audiences:

* It should be comprehensible to a non-specialist audience.
* It should avoid detail and keep the big picture in view [10].

### Purpose

As the name suggests, a strategic roadmap is a high-level document which sets out the vision for the project and elaborates its goals. It is used for a number of purposes, including the following:

* **To clarify strategy.** A strategic roadmap should outline the vision and scope of the proposed software to ensure that all stakeholders understand and expect the same thing from the project.
* **To identify priorities.** A strategic roadmap needs to make clear which of the proposed tasks and features are most important and why.
* **To monitor progress.** A strategic roadmap is used to keep track of the development of the project.

As this last purpose suggests, strategic roadmaps are kept 'live' throughout the life of a project. They are therefore subject to change in response to feedback. This is especially important to an agile approach which is constantly testing the market with new features.

### Layout

There is not an established strategic roadmap which covers all situations. They are often created as tables with bullet points and a Gantt chart, but they are also written as short narrative documents with subheadings and paragraphs. However, there are a number of features which strategic roadmaps often include. We look at those features here and provide examples and useful language for each of them.

* **Rationale**: a strategic roadmap should clearly explain what the software is meant to do and why. For example, is it addressing an issue or exploiting an opportunity? What are the benefits to the end user?
  * *Example*: the ability to use QR codes for micro-payments will benefit service industry workers, such as waiting staff and mobile delivery drivers.

- *Useful language*: this app addresses the issue of . . . , This app will differentiate itself by enabling end users to . . . , This software exploits the gap between . . .

- **Features**: while the overall vision informing a roadmap may change little, the feature list will likely develop in line with feedback. Typically, features will involve a group of themed tasks.

  - *Example*: this app will allow users to transfer micro-payments securely between accounts using mobile-to-mobile Near Field Communication.
  - *Useful language*: this app will allow X to do Y, The main purpose of this software is to . . . , The unique selling proposition of this app is . . .

- **Risks and mitigation strategies**: strategic roadmaps also often list potential risks and how they can be mitigated or minimized.

  - *Example*: take-up could be affected by perceptions of fraud. This can be minimized by advertising a user-determined limit to possible transactions.
  - *Useful language*: it is possible that. . . , In the unlikely event of. . . , Should X happen, then Y can minimize its effects.

- **Milestones**: the roadmap should include a schedule of expected milestones. This information can be used to track the progress of the project and manage client expectations.

  - *Example*: testing of NFC micro-payment feature will start at the beginning of Q2.
  - *Useful language*: at the start of . . . , no later than . . . , by the end of . . .

The milestone section normally dominates the view of the roadmap which stakeholders first see, but it should not be very detailed, or it should at least hide the detail behind dropdowns. This is because the focus of a strategic roadmap is the big picture. Detail is not required at this level.

### Technology roadmaps

Sometimes called an IT or even a technical roadmap, technology roadmaps are common across all aspects of IT departments, not just the DevOps team. They are very often used to clarify the value and use of different system aspects, helping maintain the complex tech stacks which underpin most organization's operations in the modern world. In this sense they are used to identify in-house technologies which are due to 'sunset' or be phased out, and those 'sunrise' technologies which can be phased in to upgrade them. However, in software development, a technology roadmap is slightly different and is used to offer the project team a more granular or lower-level view of development targets than the strategy roadmap.

### Audience

Used by engineering or tech leads, it is more detailed than a strategic roadmap, and uses more technical language as a result. Nevertheless, like a strategic roadmap, a technology roadmap can be used to show other non-specialist stakeholders what is scheduled in the development pipeline.

### Purpose

Technology roadmaps are used to keep track of the development schedule – what is being done as well as what has been completed and what is scheduled to be done. It therefore enables tech leads to organize and show the choices that have to be made between possibilities and resources, such as different items on the backlog of a feature wish list.

### Layout

Technology roadmaps in development are mostly graphic documents, either on Kanban boards or software. They are usually organized by swimlanes. If it's an agile project, the swimlanes can be segmented by sprint cycles. These sprints are the two-week to three-month blocks of time allocated for the development and testing of specific features, such as in Table IV. Software versions also often allow users to switch to a simple view of completed and incomplete developments, which again can be seen in Table V:

## Release roadmaps

Release roadmaps are perhaps the simplest of all the roadmaps considered here.

### Audience

Release roadmaps can be used for a variety of audiences, but if a strategy roadmap exists, then the release roadmaps are for the development team. However, as the main focus of release roadmaps is time, technical details are kept to a minimum. This makes them suitable for most audiences.

### Purpose

Release roadmaps are used to keep track of scheduled releases. Typically, this means that they chart some or all of the following information:

- The release date
- The time frame of sprints
- The release name and/or version number
- Project milestones

*Table IV* Simple technology roadmap for SDLC

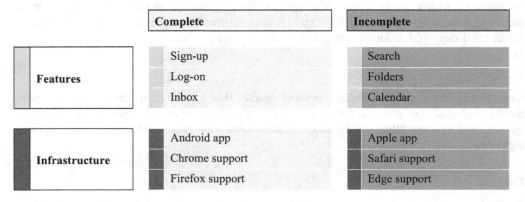

| | Complete | Incomplete |
|---|---|---|
| **Features** | Sign-up | Search |
| | Log-on | Folders |
| | Inbox | Calendar |
| **Infrastructure** | Android app | Apple app |
| | Chrome support | Safari support |
| | Firefox support | Edge support |

*Table V* Excerpt from release roadmap

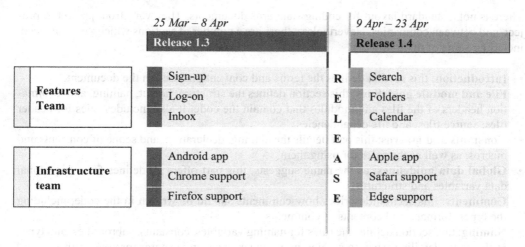

## Layout

Release roadmaps are also graphic documents, usually based on Kanban boards, but sometimes simply tables. They are generally ordered vertically by sprints and/or deadlines, and horizontally by teams and/or priorities. Table V is an example of a release roadmap in a software development cycle.

## 4.4   Other process documents

This section looks briefly at two other kinds of process documents: coding standards and working papers.

### Coding standards

Coding standards documents are written for different coding languages and different industries to ensure the safety, security, reliability, portability, testability, and maintenance of code. While these coding standards are applied across industries, individual projects also maintain their own coding standards documents.

### Audience

Coding standards documents are highly technical documents designed for engineers.

### Purpose

The purpose of a coding standards document is to ensure that the quality and consistency of the code is maintained (and is maintainable) regardless of the individual coders working on it. As such, it is a vital document for efficient collaboration among and between developer teams.

*Layout*

There is not a standard layout for coding standards documents. They vary from project to project, and between companies. Nevertheless, there are a number of sections which you may need, including some of the following:

- **Introduction**: this section defines the terms and conventions used in the document.
- **File and module guidelines**: this section defines the structure, layout, naming, and information headers of the files and modules that contain the code. It also includes rules for header files, source files, and file dependencies.
- **Constants and macros**: this part details the naming, declaration, and scope of constants and macros, as well as the rules for using them.
- **Global data guidelines**: as the name suggests, this part offers guidelines for using global data variables and structures.
- **Comments**: this section describes how comments should be written in the code, including the types, formats, and contents of comments.
- **Naming**: this section defines the rules for naming variables, constants, subroutines, and types in the code, detailing when to employ upper/lower case, underscores, prefixes, suffixes, and abbreviations in naming identifiers.

There may be other sections outlining project-specific guidelines, such as rules for comparisons, conditionals, error handling, debugging, and so on. The purpose of establishing these conventions is to improve the quality of the code. This in turn makes development cycles shorter and more productive.

### Working papers

Probably the least formalized set of documents in the SDLC process, but at the same time the most common, are the working documents of engineers [11]. These record ideas, sketches, and solutions to project issues. They are not the final documents for any project, but they do offer insights into why and how things were done. As such, it is helpful if they too employ coding standards, and are centrally stored.

## 4.5   SDLC documentation best practice

As we have seen, there are many advantages to SDLC documentation, from improving team performance through better product use to decreasing the support burden. But what are the best ways to achieve this? In this section we answer that question by looking at best practice both for writing documents and for managing them.

### SDLC document writing guidelines

There are many factors which can help improve your documentation. What they all aim to do is help the reader achieve what they want from your document and reduce the friction involved in doing so. Here are seven tips that can help:

1  **Describe the task, not the technology.** Think what your reader wants from the document and write what they need to know, not what you want to say. If your reader is an end user, for example, it is unlikely that they will want to read a detailed description of the leaf nodes

in your date tree structure. However, they will want to be able to use a certain feature in the least number of steps.

2 **Write less.** Following on from the previous point, if you make your documentation targeted, it doesn't need everything in it. This means that there's less clutter for your reader to work through, and it becomes easier to find what they want. There is a trade-off here between usability and comprehensiveness, but the balance should always be in favour of making it easy to use.

3 **Use the language your reader uses.** If you are writing for developers, use technical language. If you are writing for non-specialists, use everyday language. Both will help your reader access what they want more effectively.

4 **Create from the top.** Documents should begin with an outline. If you are looking for a city on a map, you move from the continent to the country to the area to the city. The same is true for documents. Start writing it by thinking about the overall structure, and then provide an easy way for your reader to focus in on the individual elements. Approaching the creation of your documents from the top down not only makes it easier for you to write but also makes it easier to read.

5 **Give examples.** If you are writing for a developer audience, for instance, provide code examples for likely use cases [12]. This is often easier to understand than an explanation written in text.

6 **Avoid FAQs.** Frequently Asked Questions (FAQs) are a convention for many user interfaces. However, they are more convenient for writers than for readers. If you think a reader needs to know something, put it in the section where they need to know it, not in a section where the answer is out of context. Moreover, adding FAQs leads to repeating what is already written, which leads to bloated search results for the user.

7 **Try them out.** If you write a document, go back to it a few days later and use it yourself. Is it easy to understand? Can you find what you wanted to with no problems? If so – well done!

### SDLC document management

There is little point creating great software development documents if they are not managed well. Indeed, much of the value of this documentation comes from how it is managed as much as how it is written. The following ten tips are some factors to think about when managing your documents:

1 **Documents should be updateable.** They should not be created in a once-and-for-all format [13]. This means the medium should ideally be online and not published and bound.

2 **Documents should be kept up-to-date.** Documents need to reflect the changing nature of the development they describe, whether that is the process or the product. If this does not happen, stakeholders lose trust in the software [14]. Therefore, only write documents that you are certain you can keep up-to-date.

3 **Documents should have a systematic process for keeping them up-to-date.** When a product is updated, or a process, or a role profile, or anything else, this should trigger a reminder to update the relevant document. Everyone on your team needs to know what that is and how it works.

4 **Documents should be version numbered.** When documents change because they are updated, this needs to be reflected in the document numbering. This numbering should also be tied to a date. This makes sure that everyone is using the most up-to-date version.

5   **Documents should be shareable.** One of the key benefits of software development docu-
    mentation is how it facilitates input and collaboration. In order for it to do this, documents
    need to be accessible to all the people connected with the project.
6   **Documents should be shared.** If documents are shareable, then they should be shared, not
    just with your immediate colleagues but wider stakeholders, such as neighbouring teams
    working on other parts of the project you are contributing to.
7   **Documents should have contents.** As we mentioned earlier, software documents need to
    be easily accessible. One of the features that makes them accessible is a table of contents.
8   **Documents should be hyperlinked.** To make it easier to move from one document to
    another, key references within documents should be hyperlinked. Wikis are ideal for this.
9   **Documents should have a glossary.** Many of the people reading your document will not
    know all the terms you are using. Many of the people reading your document who do know
    the terms may understand them differently. If the term is important to the project, define it
    in a glossary so that everyone understands it in the same way.
10  **Documents should be paired with source code.** Software documentation explains what
    the source code is supposed to be doing and why. It therefore makes sense to keep the two
    together so that one year from now when someone is looking at the work you have done,
    they are able to understand it because you kept it in the same place and made it easy for them
    to access it.

Many of these rules can be achieved with the use of a document management system, such as
GitHub, Document 360, or Confluence. These provide live document repositories or folders that
can be accessed by anyone in your team. They also provide templates to help you with some of
the basic tasks. However, less bespoke software, like Notion and Google Docs, offer much of
the same functionality and template options.

### Docs as code

Over the past decade or so, there has been a growing trend to adopt a *Docs as Code* approach
to software development documentation [15]. As the name implies, this involves writing documents in the same way that code is written. Amazon Web Services, for example, moved to a
Docs as Code process in 2021 [16]. Whereas previously they had created and edited documents in Word before publishing them as PDFs, they started to write and edit in ASCII, as
engineers do, before publishing in HTML using an automation script. All of this takes place on
GitHub, where developers, technical writers, and other stakeholders can collaborate and track
all changes using GitHub commits. Docs as Code is part of a broader tendency, called *DocOps
[17]*, for technical writers to work more like and with engineers.

## 4.6   Review

The following questions will help you consolidate your knowledge about writing process documents and deepen your understanding through practice.

### Reflection questions

1   What different phases would you expect in an SDLC?
2   What are two approaches to managing an SDLC, and what are their differences?

3   What is the difference between process and product documentation?
4   What is the value of documenting the SDLC process?
5   What is the purpose of a work breakdown structure?
6   When would you use a Gantt chart?
7   When would you use a Kanban board?
8   What are the purposes of a strategic roadmap?
9   Why is writing less a good principle for process documentation?
10  Why would you store process documentation with source code?

## Application tasks

1   Complete Table VI to show your understanding of different approaches to the SDLC.

*Table VI* Different approaches to the SDLC

|  | Waterfall development | Agile development |
| --- | --- | --- |
| **Time** |  | Based on sprints – short cycles of two weeks – two months |
| **Flow** | One direction |  |
| **Dependencies** | Each step follows the previous one and cannot happen without it |  |
| **Deliverables** | The project is completed at the end | The project delivers parts of the project on a regular cycle |
| **Client input** | At the beginning |  |
| **Budget** |  | Flexible |
| **Documentation** |  | Continuously revised |

2   Create a work breakdown structure for making a salad. Ensure you have at least three summary tasks, six subtasks, and 12 work tasks (this is an amazing salad!).
3   Answer these questions about Gantt charts.
   3a  What direction represents time on a Gantt chart?
   3b  What direction represents task hierarchy on a Gantt chart?
   3c  What kind of word is used to describe a task – a noun or a verb?
4   Create a Gantt chart for the salad you outlined in question 2.
5   Answer these questions about Kanban boards.
   5a  What is the minimum number of columns on a Kanban board, and what are they about?
   5b  What is a swimlane?
   5c  What is the WiP?
6   Create a Kanban board for your salad. Ensure you have at least two swimlanes.
7   You are the lead developer for a team creating an app for the software development cycle for a documentation software company. You need to produce a strategic roadmap. Write two to three sentences for each of the following sections.
   7a  Rationale:
   7b  Features:
   7c  Risks and mitigation strategies:
   7d  Milestones:

## Works cited

1   C. R. Prause and Z. Durdik, "Architectural design and documentation: Waste in agile development?," in *2012 International Conference on Software and System Process, ICSSP 2012 – Proceedings*, Piscataway, NJ, USA, 2012, pp. 130–134. https://doi.org/10.1109/ICSSP.2012.6225956

2   E. Aghajani, C. Nagy, O. L. Vega-Márquez, M. Linares-Vásquez, L. Moreno, G. Bavota, and M. Lanza, "Software documentation issues unveiled," in *2019 IEEE/ACM 41st International Conference on Software Engineering (ICSE)*, Montreal, QC, Canada, 2019, pp. 1199–1210. https://doi.org/10.1109/ICSE.2019.00122

3   E. Aghajani, C. Nagy, M. Linares-Vásquez, L. Moreno, G. Bavota, M. Lanza, and D. C. Sheherd, "Software documentation: The practitioners' perspective," in *ICSE '20: Proceedings of the ACM/IEEE 42nd International Conference on Software Engineering*, New York, NY, USA, 2020, pp. 590–601. https://doi.org/10.1145/3377811.3380405

4   J. Nawrocki, M. Jasinski, W. Bartosz, and A. Wojciechowski, "Extreme programming modified: Embrace requirements engineering practices," in *Proceedings of the IEEE International Conference on Requirements Engineering*, Essen, Germany, 2002, pp. 303–310. https://doi.org/10.1109/ICRE.2002.1048543

5   A. K. Rath and H. Mohapatra, *Fundamentals of software engineering designed to provide an insight into the software engineering concepts*. BPB Publications, 2020.

6   C. Besner and J. B. Hobbs, "The perceived value and potential contribution of project management practices to project success," *Project Management Journal*, vol. 37, no. 3, pp. 37–48, 2006.

7   G. Fernandes, S. Ward, and M. Araújo, "Identifying useful project management practices: A mixed methodology approach," *International Journal of Information Systems and Project Management*, vol. 1, no. 4, pp. 5–21, 2013.

8   J. Varajão, G. Fernandes, and H. Silva, "Most used project management tools and techniques in information systems projects," *Journal of Systems and Information Technology*, vol. 22, no. 3, pp. 225–242, 2020.

9   E. Corona and F. E. Pani, "A review of Lean-Kanban approaches in the software development," *WSEAS Transactions on Information Science and Applications*, vol. 10, no. 1, pp. 1–13, 2013.

10  R. Phaal and G. Muller, "An architectural framework for roadmapping: Towards visual strategy," *Technological Forecasting and Social Change*, vol. 76, no. 1, pp. 34–39, 2009. https://doi.org/10.1016/j.techfore.2008.03.018

11  T. Theunissen, U. Van Heesch, and P. Avgeriou, "A mapping study on documentation in continuous software development information & software technology," *Information and Software Technology*, vol. 142, pp. 1–29, 2022. https://doi.org/10.1016/j.infsof.2021.106733

12  M. P. Robillard, "What makes APIs hard to learn? Answers from developers," *IEEE Software*, vol. 26, no. 6, pp. 27–34, 2009.

13  T. Waits and J. Yankel, "Continuous system and user documentation integration," in *2014 IEEE International Professional Communication Conference (IPCC)*, Pittsburgh, PA, USA, 2014, pp. 1–5. https://doi.org/10.1109/IPCC.2014.7020385

14  I. Hadar, S. Sherman, E. Hadar, and J. J. Harrison, "Less is more: Architecture documentation for agile development," in *2013 6th International Workshop on Cooperative and Human Aspects of Software Engineering (CHASE)*, San Francisco, CA, USA, 2013. https://doi.org/10.1109/CHASE.2013.6614746

15  A. Gentle, *Docs like code write: Review, test, merge, build, deploy, repeat*, 2nd ed. Just Write Click, 2017.

16  M. R. Johnston and D. May, "Marcia Riefer Johnston & Dave May – one AWS team's move to docs as code," 2 June 2022. [Online]. Available: https://youtu.be/Cxuo3udElcE?si=vl4eFhOwi3LExyRR [Accessed 30 March 2024].

17  J. Putrino, "What is DocOps, anyway?." [Online]. Available: www.writethedocs.org/guide/doc-ops/#what-is-docops-anyway [Accessed 30 March 2024].

# Chapter 5

# System documents in SDLC

## 5.1 Introduction to product documentation

This section looks at the two kinds of product documents that are used throughout the software development cycle (SDLC).

### Two types of product documentation

The overriding purpose of product documentation is to ensure that developers, clients, management, and other stakeholders are all aligned in terms of the aims of the SDLC. To that end, product documentation describes the product being developed, how it works, and how to use it. Conventionally, such documentation is split into two types – *system* and *user* documentation:

- **System documentation** includes all the documents which describe the product being developed. It is needed *before* and *during* development to determine exactly what is worked on, or in the case of API documentation, to help others use your product in the development of theirs. Mostly aimed at developers, system documentation includes product requirement, user experience (UX), design, and quality assurance documents.
- **User documentation** includes all the documents which inform users how to operate the software. It is needed *after* development to show people, including general consumers and system administrators, what to do with what has been developed. User documentation includes installation and user guides, as well as reference and troubleshooting manuals.

In Chapter 6 we will look at user documentation, but the focus in this chapter is on system documentation.

## 5.2 Product requirement document

This section looks at product requirement documents (PRDs). A PRD is the foundation document for a software development project and, as such, it provides everyone concerned with a single source of truth. It specifies what is being made, why, and for whom.

### Audience

One of the main audiences for PRDs are engineers. In very big companies, product managers will often oversee the writing of them, but in smaller companies it will fall to management or

DOI: 10.4324/9781032647524-5

tech leads. In either case, individual developers and engineers will collaborate on it, have to read it, and ultimately action it.

Given that PRDs are a means of establishing a common vision of a project between specialists and non-specialists, everyday language is the common ground, and there is little in the way of detailed technical work.

## Purpose

Its purpose is to document the requirements of the project so that all stakeholders understand what needs to be done. These requirements are usually split between functional and non-functional requirements:

- **Functional requirements** are the product features which have to be created so the software can perform its task. For example, a functional requirement of many apps is that it automatically sends an email or text message for account verification.
- **Non-functional requirements** are descriptions of how the software should perform. For example, it is a requirement of apps with verification systems that they send the email or text message immediately and should be able to do so even if 1000s of users are on the system at the same time.

In sum, functional requirements describe product *features* which are *essential*, whereas non-functional requirements describe product *properties* which are only *desirable*. In terms of how they are usually documented within a PRD, functional requirements are determined by use cases, while non-functional requirements are determined by descriptions of attributes.

## Layout

Like most IT documents, PRDs are written for accessibility using section headings and sub-headings with small paragraphs, where necessary, and diagrams where optimal. There are no set formulas [1], but PRDs usually include some or all of the following sections:

### Key information

The beginning of the document should identify the name of the product, the document version, the release dates, and the key stakeholders in a short table, such as Table I. These stakeholders may include the product manager, the designer, the development team, tech, and QA leads, depending on the size of the project. Very often, a product will be for another team in the same company, and so they may be a visible stakeholder too. What this section does is enable people to verify what the product documentation is, what version it is, and who is responsible for which parts.

### Problem

This is variously identified as the *challenge*, or the *issue*, but it refers to an overview of what it is the product is being designed to help with. Making this the first section in your PRD not only places the focus on where the product idea comes from, but it also does so from the point of view of customer need. As we will see throughout the document, a good PRD is not about *what the company wants* to make, but about *what the customer needs*.

*Table I* Example of key information section in PRD

| Product Title | |
|---|---|
| Last updated on dd/mm/yyyy | |
| Target release | 2.1 |
| Document owner | @xchen |
| Designer | @bwesley |
| Developers | @xchen; @jjacson; @yyijun |
| Quality Assurance | @bweiwei |
| Doc version | 1.4 |

The simplest way to begin your PRD with a definable problem is to answer some questions which start generally and become more focused. For example, the following four questions do this:

- What is the problem?
- How is the problem measured?
- What can client/user not do?
- What do they want to be able to do?

Answering these questions will help you to identify more clearly the overall issue that your programme is designed to solve. We can see an example in Table II of a PRD which begins with a problem regarding a food delivery service app.

Notice that the problem is defined as a general issue before it is specified more precisely by measurements or metrics. These metrics help the team articulate the specific challenges causing the problem, challenges which can then be articulated as needs. The needs form the basis of the user stories.

## Solution

This section outlines the proposed solution to the problem. This does not require technical detail but will offer a broad overview of how the issue can be resolved. For example, the solution to service delays might be *managing customer expectations with a countdown timer informed by a GPS route-mapping software.*

## Objectives

This is where the objectives of the project are defined, usually in a bulleted list. Ideally, the objectives should be written in a measurable form. Take a look at the following example:

The aim of this project is to

- Increase customer return rate by 25% to achieve industry standard within 12 months.
- Decrease delivery delay complaints by 50% within 12 months.
- Decrease complaints per user ratio to industry standard of 1:58 within 12 months.

Here the aims are aligned with the issue metrics outlined in the *Problem* section. They are written in a measurable, time-constrained way so that the success of the project can be clearly understood by all the stakeholders.

*Table II* Example of problem section in PRD

| Issue | A food delivery company receives unacceptable volume of complaints about delivery delays. |
|---|---|
| **Issue metrics** | • Customer return rate is 20% below industry standard.<br>• Delivery delay complaints constitute 95% of all complaints.<br>• Complaints per user ratio is 1:14 compared to industry standard of 1:58. |
| **Challenge** | End users are not able to make informed decisions about what to order:<br>• User cannot see estimated time for delivery *before* ordering.<br>End users' delivery time expectations are not managed while waiting for delivery:<br>• User cannot see delivery time *after* ordering. |
| **Need** | • User needs to see estimated time for delivery *before* ordering.<br>• User needs to see delivery time *after* ordering. |

### User stories

User stories are the engine of the SDLC. They are the specific accounts of the user needs outlined in the *Problem* section. They do not refer to the features or implementation of them, but simply the story of the end user's want from their point of view.

They are typically written using the following formula:

**As a** [user type], **I want to** [perform a function] **so that** [benefit].

As we can see here, user stories are short, but they include three vital pieces of information: the user persona (the *who*), the operation they want to complete (the *what*), and the outcome they would like to achieve (the *why*). Let's look at some concrete examples:

- **As a** bank customer, **I want to** see recent credit card transactions **so that** I know what my card is being used for.
- **As a** racing game user, **I want to** see my racing position on-screen at all times **so that** I know where I am relative to the NPCs.
- **As a** food delivery app user, **I want to** know how long a delivery will take to arrive before I order **so that** I can order from a restaurant which delivers within an acceptable time frame.

None of these are written in technical language and so are readily understandable to all stakeholders. This is important as this fairly informal approach to determining product features is a collaborative one. User stories are as much about including stakeholders in a conversation with development teams, as they are about informing the design of the feature in a sprint.

There is no single dominant template in user stories [2], but good ones are often written using the *INVEST* approach. This is an acronym which stands for a set of ideal attributes for each user story:

- *Independent* – user stories should not be dependent on other user stories so that they can be sequenced in any order.
- *Negotiable* – user stories should allow for different approaches to meet the needs of the user.
- *Valuable* – user stories need to add value to the user.

- *Estimable* – user stories must provide an indication of the resources required to meet their needs.
- *Small* – user stories need to point to a solution which can be created within a sprint cycle.
- *Testable* – user stories should point towards at least one acceptance criterion which can confirm its completion.

These attributes are not always easy to achieve in every user story, but they are helpful as guidelines which can be used to help make user stories more practical for the development team.

User stories are also important for planning poker, where development teams meet to assign points to stories in order to estimate the effort required to complete them. This is done based on the complexity of the task, the amount of work involved, and the level of risk and uncertainty. Stories are then given a relative value using t-shirt sizes or numbers from the Fibonacci sequence in order to determine how many resources they need to be given to make the feature that answers the user want. For example, a feature requiring two engineers for two weeks might be a medium t-shirt, but one requiring four engineers for a month could be an extra-large. It is therefore crucial that user stories are carefully crafted.

### Epics

If a user story does not meet the INVEST criteria, it could well be that it is still what is known as an *epic*. Epics are essentially large user stories [3]. In fact, they usually contain multiple user stories and are considered too big for a sprint cycle. They are generally written in the same way as user stories. Let us look at the following example:

- **As a** car buyer, **I want** to be able to get a good overview of all secondhand cars **so I** can understand what a good price to pay is for the model I choose, and I can buy one with peace of mind.
  This is a useful story and one that presents developers with an opportunity, but if we consider the INVEST criteria, it is not estimable, small, or testable in its current format. It would therefore need to be broken down into smaller, testable, and estimable user stories, such as these:
- **As a** car buyer, **I want** to be able to compare durability across all available car providers **so I** can easily see which ones are of good value to me.
- **As a** car buyer, **I want** to be able filter cars according to engine size, usage, model etc., **so I** can compare similar cars more easily.
- **As a** car buyer, **I want** to get relevant search results based on key parameters **so I** can know which individual offer is best for me.

### Acceptance criteria

However good the user stories are, they do not offer enough detail for developers to be able to create features that meet the wants and needs of the user. For example, a user story may state that a user wants to be able to search through different restaurant cuisines on a food delivery app. A developer may make that possible with a search bar, but what the client wanted was for the end user to make a choice from available options. That is a very different experience.

To avoid such issues and to ensure a consistency of understanding between all stakeholders, user stories have to be converted into *acceptance criteria*. Acceptance criteria are descriptions of exactly how functionality is provided to users to fulfil their needs. They usually take one of two forms – rule-oriented or scenario-oriented. Let us look at each in turn.

**Rule-oriented acceptance criteria** are essentially descriptions of the rules which govern a product. These tend to be used when the development team does not require specifics or if system-level functionality is being referred to. They usually take the form of a bulleted list, such as this example:

- Search bar is placed in the middle of the page.
- Search bar contains placeholder text *What do you want to find?*
- Placeholder text disappears when user starts to type in search bar.
- Search starts when user presses either *Search* or *Top Result*.

While such criteria are useful in describing systems, as you can see these are not so much about what the user does. For that we need to use a scenario-oriented acceptance criteria.

**Scenario-oriented acceptance criteria** are descriptions of what the user does in each scenario. They are usually written in a given-when-then formula, where *given* refers to the preconditions, *when* refers to the user action, and *then* refers to the result. In Table III we can see what each component of a scenario-oriented acceptance criterion means, and an example based on our earlier user story about wanting a delivery estimate as a food delivery app user.

As you can see here, there is a lot more detail in the acceptance criteria than in the user story. While user stories are acceptable for the backlog in an agile project, they should only be put into a sprint once all the stakeholders have agreed on the acceptance criteria.

*Table III* Example of a scenario-oriented acceptance criterion

| Scenario-oriented acceptance criteria | | |
|---|---|---|
| **User story** | As a food delivery app user, I want to know how long a delivery will take to arrive before I order so that I can order from a restaurant which delivers within an acceptable time frame. | |
| **Components** | *Meaning* | *Example* |
| SCENARIO | The name of the described behaviour | Providing estimated delivery times |
| GIVEN | The start of the scenario | User logs into their account AND User keeps location up-to-date |
| WHEN | The action performed by the user | User clicks on a menu item AND User clicks on basket |
| THEN | The result of the action | User sees estimated delivery time displayed above the basket |
| AND | The continuation of any of the aforementioned | |

There are a number of attributes which good scenario-based acceptance criteria tend to have, including the following:

- *Specific* – the specificity of the acceptance criteria helps all aspects of product development. They help define the test parameters of the quality assurance tests, for example, as well aid the creation of mock-ups for the UX designer. However, do not be overly specific in how a scenario should be manifest. Describe the components, not the feature.
- *Concise* – be concise in your descriptions. Use several simple sentences rather than one complex one.
- *Non-technical* – avoid technical jargon so that all stakeholders can understand the criteria.
- *Active* – write using the active voice, not the passive voice so that the user is always doing something. For example, do not write *The basket is clicked on* (passive) but *The user clicks on the basket* (active). Keep *who* is doing something central to your criteria.
- *Positive* – it is best to avoid using negative expressions, such as *not*, as these can lead to confusion.

## User experience

This section describes how the user interacts with the product. It can contain a bulleted list of the different stages of user interaction, but it usually consists either of wire frames and mock-ups or a link to them.

## Scope

This section outlines what the project will *not* be doing. It is always helpful to delimit a project so that it does not suffer from *mission creep*. While the agile approach to development and testing means a project may evolve, adding features that were not originally thought of, that is not the same as expanding unnecessarily.

**Key vocabulary**

*Mission creep* – this term refers to when a project extends beyond the goals it was originally given. Good documentation helps to stop mission creep.

A simple bulleted list is usually enough to outline the boundaries of a project in a useful way. Here are some examples.

- This app will not include a feature to allow micro-transactions in non-native currencies.
- This app will not include functionality to permit users to direct micro-transactions to non-native accounts.
- This software will not incorporate support for Android users.

As you can see, these are not detailed but provide recognizable limits to the direction of the development. The use of such a list is not only in keeping a project focused on what it is doing now, but also on possible areas for development following the current SDLC. For example, the next SDLC might look at developing support for Android.

*Questions*

This is a live section where significant issues are documented along with the answers that were decided upon. This enables all stakeholders to understand when and why challenges were dealt with.

The *Questions* section is usually the last part of a PRD; however, it is worth emphasizing again that PRDs have different layouts and different lengths – sometimes a single page long and sometimes five to ten pages. These differences depend on the scale of the project and whether or not the PRD incorporates the functions of some of the other documentation we will look at in this chapter.

## 5.3   Software requirements specification document

A software requirements specification (SRS) document often comes from, and has crossover with, the PRD. In broad terms, it is a more technical document than the PRD and is more concerned with details. If the PRD focuses on the *why*, the SRS attends to the *how*.

### Audience

An SRS document is generally written for the development and testing team, as well as maintenance, rather than other less specialist stakeholders for whom the PRD is a more suitable document.

### Purpose

The purpose of an SRS is to provide developers with a single source of truth for the software being worked on. It is hard to overstate the value of this. It is estimated that poor documentation is the single greatest cause of project failures [4]. One famous example of this was the Mars Climate Orbiter built by NASA in 1998, where separate software teams used newton-seconds instead of pound-force seconds, causing the $125 million spacecraft to disappear [5]. These are the kind of assumptions which an SRS is designed to reveal.

### Layout

As with PRDs, SRS documents follow different layouts, despite attempts to systematize them by the IEEE and others. In what follows we look at some of the features which you can expect to find in an SRS. These features are often supplemented by others, especially if the expected audience is larger than the development team.

*Table IV* Example of question section

| Date | Question | Raised by | Answer | Answered by |
|------|----------|-----------|--------|-------------|
| 2.2.24 | Should we enable support for multiple bank accounts? | AJ | No, as per discussion, this will be enabled in a later iteration. | ZZ |
| 8.2.24 | Should we introduce QR code functionality? | QZ | No, security protocols make this too complicated for a single sprint. | ZZ |

## Introduction

Typically, the introduction contains the following elements:

- **Purpose**: this is a very brief outline of the purpose of the SRS.
- **Intended audience**: this details who the intended readers of the document are within the organization. If the SRS is designed for those outside of the development team, particularly non-specialists, then this will impact the level of technicality that can be used.
- **Project scope**: this contains a brief description of the software, including its name, what the software is intended to do, and how it relates to the needs of the business, includi n g its aims and objectives.
- **Definitions, acronyms, and abbreviations**: this section defines key terms and explains acronyms and abbreviations used throughout the document, as well as the meaning of any typographical conventions.
- **References**: this section lists the references to reports and external links used in the document.

## Overall description

The second part of an SRS is usually dedicated to describing the software under development.

- **Product perspective**: this section describes the context of the software, whether it is independent or part of a larger system, and if so, how.
- **Product functions**: this part lists the main product features.
- **User classes and characteristics**: this section differentiates between users according to relevant factors, such as likely frequency and use, indicating possible characteristics, including level of technical expertise and experience.
- **Operating environment and constraints**: this section details the hardware and operating system in which the software will work.
- **Design and implementation constraints**: this part identifies any restrictions, such as memory, security issues, or interfaces, which might restrict the software design.
- **User documents**: this part lists the user documents that will accompany the completed software.
- **Assumptions and dependencies**: this outlines any assumptions underlying the software development, like the continued existence of an operating system, or any dependencies, such as components of the software.

## System features and requirements

The third part of an SRS tends to describe the software under development.

- **Features**: describe the features of the software here and indicate their development priority.
- **Functional requirements**: identify the functional requirements for the software features, including defined responses to user error and invalid inputs.
- **Non-functional requirements**: set the performance, safety, and security parameters here, as well as detailing any other kind of characteristics that user stories indicate would be valued, such as localization and reliability.

## 5.4  UX design documentation

User experience (UX) design documentation is usually created by a specialist designer. However, as UX refers to any part of an end user's interaction with a business' software, this is not an area that can be ignored by any part of the development team, particularly as UX design teams have come to adopt the same methodologies as agile developers [6]. Indeed, leading UX design software, such as Figma, includes development team access as acknowledgement of the all-inclusive nature of UX.

### Audience

While a UX design document is intended for all the current stakeholders on a project, it has particular relevance for the development team. It is also designed for future stakeholders who will need to get up to speed quickly on a project that they were not initially part of. In this regard, a UX design document should always be written for someone who is coming to the project for the first time.

### Purpose

Like the other documentation mentioned here, UX design documents are designed to maintain a single vision for a project. The documents are intended to be live, reflecting the ongoing and changeable nature of development but also future-oriented in so far as they provide a plan for moving forward. They are also intended to be legacy documents so the reasoning behind decisions can be understood by future stakeholders. In all three aspects, they are designed to improve collaboration as they are open to all stakeholders and written in non-technical language wherever possible.

### Layout

A UX design document will usually contain many of the elements of a PRD, such as user scenarios and stories. Much of the information is represented graphically. Two important examples of this are user flows and wireframes.

#### User flows

User flows, or interaction flowcharts, depict the journey of a user through a programme from the beginning to the end of a particular transaction. They show the route the user has to take and what kind of issues they face on the way to completing what they want to do. This enables designs to be modified to maximize the ease of use.

Typically, user flows are represented graphically with text limited to descriptions of the nodes in the flow. These nodes usually take the form of different shapes which represent different parts of the flow, as we can see in Figure 5.1:

**Rectangular oval** – beginning or end. These mark the beginning and end of processes, such as opening the homepage of an app.

**Rectangular square** – action. These are the most common symbols in user flows. They designate steps in a process, such as creating an account.

**Diamond** – decision. Diamonds are most often used to mark points in the user flow where the user is given options and the flow splits into two, such as when users are asked to confirm data inputs.

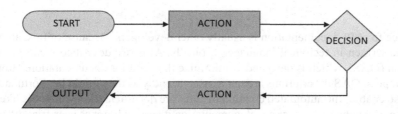

*Figure 5.1* Main types of symbol used in user flows

**Parallelogram** – input or output. The most common use for parallelograms is where users input data, such as contact information in profiles.

**Arrow** – direction. These show the flow from one node to another and are the most common features of user flows.

Typically, user flows depict many actions and decisions which create a whole series of branching processes. Portraying these branches in this simplified graphical way helps developers to better understand the experience of the end user and what may help or hinder their use of an app.

### Wireframes

Wireframes are schema that show what elements should be in the user interface (UI) and where they should be. They do not show what the elements should look like, which is what you see in a mock-up, nor how the UI feels, which is what is shown in the final stage of a prototype. Rather they are *low fidelity* and focus on the information architecture – how information and tasks are organized and ordered in a piece of software (as can be seen in *site maps* on most websites).

As with user flows, wireframes are primarily graphic in nature. Text is limited to the text found on the UI itself. Also similar to user flows, there is a convention for using shapes to portray the final product. However, these shapes are the screens of the graphical user interface, such as the phone or laptop screen, and the shapes of the outlines of the intended buttons and displays. Indeed, sometimes the user flows and wireframes are combined to create so-called screen flows.

> **Key vocabulary**
>
> *Low fidelity* – this term is used to describe a design with minimal detail. Wireframes are intended to be low fidelity so that the focus is on the organization of elements, not what the elements look like.

## 5.5   API documentation

Application programming interface (API) is software that enables software to communicate with other software for the purpose of data input or output or other functionality. The API economy, which leverages the value of APIs to develop new businesses or make existing ones more efficient by using API mash-ups, is a key part of software development, so it is therefore vital that the documentation which allows this is usable.

## Audience

The audience for API documentation is usually other developers. Because of this, an API document can be written in technical language. Typically, APIs are described using the OpenAPI Specification (OAS), which is designed to minimize the need for documentation. However, for many developers, OAS descriptions are not as user-friendly as they might be. Fortunately, using the OAS also enables the automated creation of software documentation with tools like Swagger and Postman. Even then, however, the documentation it creates is not as user-friendly as it might be and can still be tweaked to meet the needs of all possible users.

In this regard, it is worth remembering that APIs are often intended to be used by people with very little experience or by people who are not developers but who need to make quick decisions about integrations, such as CTOs. The only way to meet the needs of a diverse readership with different technical know-how on such a technical subject matter is to write for the least technical audience. Some of the most successful APIs, such as Google's, make the DX a key focus and therefore have documentation that is written in almost non-technical English.

**Key vocabulary**

*DX* – this term is shorthand for *developer experience*. It references UX by analogy but making developers the users. It highlights the importance of making APIs developer-friendly because of the income they generate.

## Purpose

The main purpose of API documentation is to make it possible for other developers to use it. One study found that 78% of API users learned how to use them by reading the documentation [7]. It provides the procedural know-how to enable developers to integrate it with their own software. Good documentation is particularly important for REST APIs which are not standardized, unlike, for example, SOAP APIs. They are therefore all individual enough to require case-specific documentation. Rather than forcing engineers to trawl through stack overflow looking for solutions, good documentation provides users with the answers to all their most common needs in a readily accessible fashion. This underlines the real value of API documentation – to make an API attractive to developers to use [8]. In an important sense, API documentation is the UI for the API.

## Layout

As a rule, reference topics dominate API documentation. This is best written using the OAS with apps likes Swagger or Postman. However, less technical sections like overviews, getting started, and authorization sections are still best written individually to meet the needs of the widest possible audience. We will focus on these sections as they can account for up to 50% of your final document.

### Overview or about

This section describes what the API does and how it can be helpful to its users. It may include any of the following parts:

**Purpose**: what issues or problems does the API address? This can be a brief statement. For example, *This API allows developers to add real-time news data to their applications.*

**Features and capabilities**: what are the core features the API offers? This can be written as a list, such as the following:

- Data retrieval for 78 countries
- Hourly updates
- Support for newsfeed-based notifications

**Audience**: who are the primary users or developers the API is intended for? This will allow possible users to identify if the API is likely to be useful to them. This can also be a list, for example,

- Mobile app developers
- Streamers
- Stock traders

**Related Services or APIs**: is the API part of a product suite? If so, this should be mentioned here. For example, a news API might belong to a suite that includes APIs for sport results, stock updates, and currency trading.

**High-level Architecture**: how does the API interact with other components? This might include showing in a diagram where the data comes from, how it's processed, and how it's delivered through the API.

**Fair Usage**: what are the rate limits and throttling policies? Developers and other stakeholders need to be advised upfront about what constitutes fair usage in order to determine if the API if appropriate for their needs.

**Terminology**: is there any unique terminology used with the API? If so, define it here.

**Version Information**: does the documentation correspond to a specific version of the API? If so, mention it here. In case a developer is looking for another version, provide information on where documentation can be found for other versions.

**Feedback and Contribution**: do you want to receive feedback on the documentation or the API itself? If so, let readers know where they can send it. If it is an open-source project, you might also outline how developers can contribute.

### Authentication and authorization

This section describes how users can get authorized access to the API, and the different kinds of authentication it is possible to use. It may include the following sections:

**Sign-up and Registration Process**: where can users sign up or register for the service? This is usually written in the form of step-by-step instructions.

**Generation of Keys/Tokens**: how users can generate, retrieve, or reset their API keys or tokens once registered? This is also typically written as a set of instructions.

**Safe Usage**: how can keys be used safely? Given the need to communicate with all levels of audience, best practice includes advising users on how to use the API securely. This might include information about OAuth Flows.

### Status and error codes

HTTP status codes, such as the infamous 404, need to be listed in the documentation. This is particularly the case for any error codes that are unique to your API. However, while it may be assumed that other developers will know what common codes mean, it is still good practice to indicate what they are and, more specifically, what is likely to cause them in your API [8]. This will aid troubleshooting for users, improve the DX, and make it more likely that your product will be used.

It is best practice, then, to list codes with their associated text, meaning, and a description, as in Table V.

*Table V* Example of status and error codes

| Code | Text | Meaning |
|------|------|---------|
| 200 | OK | Great success! |
| 400 | Bad request | This is an invalid request. It is likely that the request is not authorized or that it used the wrong query type. |

### Quick reference guides

This section provides instructions on how to perform some of the more common tasks with an API. This is where you distil the essence of the API into one or two pages. If you have a limited number of endpoints, you could list them all with brief descriptions and links to the more detailed reference for each. This gives developers the opportunity to look through all the endpoints with a search function, as well as to get an overview of an API and how it is set out.

### References

The References section serves as a comprehensive guide, offering in-depth details about each endpoint of the API. This includes specifics about request and response parameters, along with illustrative examples of requests and responses. Unlike the Guides section, the References is written in a more succinct and technical manner, catering to developers who have prior knowledge of the API and use it as a quick reference tool.

Typically, the References section will include some of the following:

- Object or resource description
- Endpoints and methods
- Request and response parameters
- Request examples
- Response examples

**Object or resource descriptions** are usually very brief one or two sentence descriptions of a resource or of objects collected under a resource, with a tabulated list of related **endpoints** and **methods**. The methods are normally single word action verbs, such as GET, POST, PUT, PATCH, and DELETE. They are written in small block capitals and, where possible, in a colour which is systematically used throughout the documentation. In Table VI, the resource is *Customers* and the associated endpoints and methods are listed in shaded highlighting to make it easier to find.

## Customers

These objects are customers of your business. Use customers to track profiles and charges.

**Parameters**, such as header, path, and query string parameters, are often written as tables. These tables, such as Table VII, can list the name of the parameter, the type (e.g., string, integer, Boolean), the optionality, the meaning, and depending on the type, the value range.

**Request and response examples** are written in code. If it is a REST API, then these are language agnostic and the ability of developers to send an http request in their programme's language of choice is often assumed. However, better API documentation will offer developers a menu of language requests which they can copy directly into their own software. This is an example of how the value of documentation can be improved by taking into account audience needs.

## 5.6   Quality assurance documentation

Quality assurance (QA) plays a vital role in the SDLC. A failure to ensure QA leads to a general failure with your product. Proper documentation is the key to this as it is what enables developers and testers to maintain successful QA regimes.

### Audience

There are many types of QA documentation and no absolute consensus on their form and content [9]. However, one way to think about them is how they can be ranked in terms of abstraction, with higher levels dealing with policy, moving through procedural approaches, to unique cases. The different document types can be seen in Table VIII.

As we can see in Table VIII, QA document types become more granular in their objectives, the more narrowly they focus on parts of the SDLC. At the same time, the audience also becomes more distinct. Generally speaking, the audience for QA documentation falls into two camps. The first group are developers and/or testers who will need the documentation in order to execute the testing plan. The second group is all other stakeholders. They do not need to execute the QA

*Table VI* Example of resource description

| Endpoints | |
| --- | --- |
| Post | /v1/customers |
| Get | /v1/customers/:id |
| Delete | /v1/customers/:id |

*Table VII* Example of parameter table

| Name | Required/Optional | Type |
| --- | --- | --- |
| Address line 1 | Required | String |
| Block number | Optional | Integer |

*Table VIII* Types of quality assurance documentation

| Document type | Objective | About | Audience |
|---|---|---|---|
| **Test policy** | Defines testing standards and criteria | Whole organization | All stakeholders |
| **Quality management plan** | Defines deliverables and roles | Project | |
| **Test strategy** | Defines testing objectives and limitations | Product | Developers and/or QA testers |
| **Test plan** | Defines test items, pass criteria, and schedule | Individual sprint | |
| **Test case** | Defines test case ID, results, and status | Individual feature | |

procedures, but they need to know they are there. For example, external customers will want to know that you have a test plan in place so that they can trust the final product will not be buggy.

## Purpose

The purpose of quality assurance documentation is to ensure an organization delivers a high-quality product that is not spoiled by bugs and other flaws which impede its performance and therefore its viability in the marketplace. QA and testing can happen without documentation, but it will not be systematic, it will not be transparent, and it will not be focused. It is the documentation which ensures all of these things.

It does this by establishing standards for testing, determining how testing should be done, when it should be done, and who should do it. The purpose of this is to make sure the right testing happens when it should.

## Layout

There is a trend toward consolidation in documentation in software development in order to keep things simple. This means that the functions of these different documents are often merged or overlap. This is often the case with test plans which can have project-wide scopes or only be viable for individual sprints. In this section we are going to look at the fuller version of a test plan, as well as at test cases. These are the documents you are most likely to work with early in your career.

### Test plan

A test plan is an outline of what needs to be tested, how, when, and by whom. It usually lasts for the length of a sprint or longer, depending on the project management approach. If it lasts longer, it takes on the features of the quality management plan and is less subject to revision.

As the kind of document used across all types of software development, test plans come in many different formats, and the best ones exceed those based only on feature lists [10]. One of the most universally accepted ones is the ISO/IEC/IEEE 29119-3:2021 (which replaced the IEEE 829 test plan standard) [11]. As we explore further in Chapter 8, the IEEE (Institute of Electrical and Electronics Engineers) provide formatting, referencing, and other standards for the IT industry. The ISO/IEC/IEEE 29119-3:2021 test plan standard contains the following sections:

# 1 Identificatory information

- This information should go on the frontispiece of your document.
- It should include a unique identifier which makes it easy to recognize the document.
- It might also include the title, the date, and document status (*Draft*, *In Review*, *Approved*, etc.).
- There should also be reference to the version history. It can be tabulated as in Table IX.

*Table IX* Example version history table

| Date | Version | Description |
|------|---------|-------------|
| 13/04/23 | BA23APP.5 | Updated test items to include account change status |
| 12/02/23 | BA23APP.4 | Updated test items to include login |

# 2 Introduction

This section outlines the areas the document covers and makes clear what it doesn't. It also provides a simple list of documents and links to repositories which may prove helpful, and lists any acronyms, abbreviations, or unusual technical terms the document uses with their meanings.

# 3 Context of testing

This section should describe what the test items are, what features are being tested, as well as what is not, and the types of tests being conducted.

# 4 Assumptions and constraints

This section outlines any cost, time, or contractual constraints, as well as the test policy and regulatory requirements that need to be met.

# 5 Stakeholders

This section needs to list the project's stakeholders and their responsibilities. One of the most common ways to do that is by using a *RACI* matrix. RACI stands for responsible, accountable, consulted, informed. Stakeholders are plotted onto a table which illustrates their function in relation to the different parts, as we can see in Table X.

*Table X* RACI matrix example

| Task | Project manager | Lead developer | Developer | UX designer | QA tester |
|------|-----------------|----------------|-----------|-------------|-----------|
| **Define project requirements** | A | C | I | C | I |
| **Create design mock-ups** | A | I | I | R | I |
| **Develop feature** | C | A | R | C | I |
| **Conduct code review** | C | A | C | I | R |

In Table X, responsibility and accountability for each stage of the project are clearly defined:

- **A** = Accountable: the person who must sign off or approve when the task is complete.
- **R** = Responsible: the person who does the work to complete the task.
- **C** = Consulted: people whose input is sought; usually, two-way communication.
- **I** = Informed: people who are kept up-to-date on progress; typically one-way communication.

**6  Testing communication**

This part describes the particular chain of communication between the testing team and the rest of the project team.

**7  Risk register**

This is a list of the product and project risks.

**8  Test strategy**

This part of the test plan details the testing approach, the level and type of the texts, the criteria for test completion, as well as the type of data that might be collected, and how the process and the testing team relate to the organization as a whole.

**9  Testing activities and estimates**

This is a list of all the test activities with their associated timelines and budgets.

**10  Staffing**

This section should provide a list of required personnel for the testing, including their roles and responsibilities in a RACI matrix. It should also highlight any expected staffing short-falls that need to be addressed.

**11  Schedule**

The schedule outlines the key testing milestones and expected dates for deliverables.

*Test cases*

While a test plan covers the overall testing approach of a project, a test case provides step-by-step instructions on how to test individual features or parts of a project. This is usually presented in a table, such as Table XI.

*Table XI* Example test case

| Project ID | BA23APP.4 | **Test case ID** | | | BA23APP.4/12 |
|---|---|---|---|---|---|
| **Test case description** | Test login page | **Test priority** | | | High |
| **Precondition** | Have a valid user | **Date** | | 03/03/2025 | |

**Test steps**

| Step Number | Action | Inputs | Expected Result | Actual result | Status |
|---|---|---|---|---|---|
| 1 | Navigate to login page | Login page address | Login page should exist at correct address | Login page existed at correct address | PASS |
| 2 | Input valid username | testid@ gmail. com | Page should present with two boxes for user-name and password | Page presented with two boxes for user-name and password | PASS |
| 3 | Input valid password | tesT1234 | User should be able to login | User was able to login | PASS |

Test cases usually include the following information:

- **Identifiers** – these can include the project ID and the specific test ID.
- **Description** – this describes what is being tested.
- **Preconditions** – this section outlines what assumptions the test needs for it to be effective.
- **Steps** and **Actions** – this part details the step-by-step procedure followed in the test.
- **Inputs** – this section identifies what was inputted at each stage and is a crucial part of ensuring the validity and iterability of the test.
- **Expected** and **Actual Results** – this is where you describe what should have happened and compare it to what did.
- **Status** – this is usually a simple pass/fail.

## 5.7  Language focus in system documentation

### Key grammar: grammatical clarity I

Advanced writers can sometimes express relationships of time as relationships of logic. Instead of using terms like *before* or *after*, they use terms like *given* or *with*. Compare these examples:

**Temporal**: before you can use the API, you need to obtain an API key.
**Logical**: given an API key, you can use the API.
**Temporal**: after the software update, performance improvements are noticeable.
**Logical**: with the software update, users will notice performance improvements.

In the logical version of each sentence, the sequence of events is not immediately clear. It is therefore more helpful to readers if relationships of time are expressed with time words, rather than hidden in logical expressions.

### Key grammar: grammatical clarity II

Advanced writers can also sometimes express relationships of cause and effect as relationships of logic. Instead of using terms like *due to* or *because*, they use terms like *given* or *with*. This makes the writing more abstract. Compare these examples:

**Causal**: because of its advanced encryption, the system is secure.
**Logical**: with advanced encryption, the system ensures security.
**Causal**: the application crashes due to memory leaks.
**Logical**: in the presence of memory leaks, the application experiences crashes.

In the logical version of each sentence, the cause is not clear. Be careful when writing that you do not hide causes by using logical expressions. They make it harder for the reader to follow the meaning.

### Key grammar: tenses

Use the present simple to describe the normal actions of an API, such as *It returns a list of users*.

Use the future simple to describe an action after an API call is made, such as *The API will then update the user's profile*.

## 5.8   Review

The following questions will help you consolidate your knowledge about system documents and deepen your understanding through practice.

### Reflection questions

1  What are the main differences in audience and purpose between system and user documentation?
2  What is the function of a product requirement document?
3  What is the difference between how functional and non-functional requirements are represented in documentation?
4  Why are user stories the driver of the software development lifecycle?
5  What is the difference between a product requirement document and a software requirements specification document?
6  What is the value of a UX design document?
7  What is the difference in the way that user stories and user flows approach the user?
8  How does a document with low fidelity diagrams help the developer?
9  Why is it important to consider DX when making API documentation?
10  How is API documentation the UI for the API?

### Application tasks

1    Look at the following problem section from a PRD in Table XII. It describes an issue for a calendar app development team which is losing market share to a more innovative rival.

*Table XII*  Problem section from PRD

| Project problem | |
| --- | --- |
| **Issue** | A calendar app is losing market share to a rival app which has leveraged APIs from online meeting providers to enable integrated appointment-making. |
| **Issue metrics** | • Customer<br>•<br>• |
| **Challenge** | End users are not able to<br>• |
| **Need** | • User needs to<br>•<br>• |

1a  How can the general issue be specified more precisely by measurements or metrics in the second part of the table? Think of appropriate metrics and complete that section.
1b  Using the metrics you created as a base, describe the specific challenges causing the problem in the third part of the table.

1c   Turn these challenges into needs in the final section.
1d   Use the metrics from the table to write two objectives for the project to tackle the issue in
     Table XIII. Begin it as follows:

*Table XIII* Project objectives

| Project objectives | |
| --- | --- |
| **The aim of this project is to**: | • |
| | • |
| | • |

1e   Now try to put the needs of the user into the form of three user stories in Table XIV. Try to
     do it using the traditional formula:

*Table XIV* User stories

| User stories | |
| --- | --- |
| **User story 1** | As a |
| **User story 2** | As a |
| **User story 3** | As a |

   **As a** [user type], **I want to** [perform a function] **so that** [benefit].

*Table XV* INVEST criteria

| INVEST test | | | |
| --- | --- | --- | --- |
| *Criteria* | *User Story 1 yes/no* | *User Story 2 yes/no* | *User Story 3 yes/no* |
| **Is it independent?** | | | |
| **Is it negotiable?** | | | |
| **Is it valuable?** | | | |
| **Is it estimable?** | | | |
| **Is it small?** | | | |
| **Is it testable?** | | | |
| **Does it pass the INVEST test?** | | | |

1f   It is now time to test your user stories using the INVEST criteria. Complete Table XV to see
     how usable your user stories are.
     If your user stories fail the INVEST test, how could they be rewritten to make them better?
1g   Once you have established your user stories, you can create more detailed acceptance
     criteria. In this instance, try and develop scenario-oriented acceptance criteria for one of
     your user stories. Write it using the *given-when-then* formula, where *given* refers to the

preconditions, *when* refers to the user action, and *then* refers to the result. Do this by completing Table XVI.

*Table XVI* Scenario-oriented acceptance criteria

| Scenario-oriented acceptance criteria | | |
| --- | --- | --- |
| *User story* | | |
| Components | Meaning | Example |
| SCENARIO | The name of the described behaviour | |
| GIVEN | The start of the scenario | |
| WHEN | The action performed by the user | |
| THEN | The result of the action | |
| AND | The continuation of any of the aforementioned | |

**1h** Finally, in this mini-PRD, it is necessary to restrict the scope of the project. Complete Table XVII with three delimitations

*Table XVII* Project scope

| Project Scope |
| --- |
| **This app will not include** |
| |

**2a** Rewrite these sentences to make the sequence of events clearer. Use terms like *before* or *after* to alter the relationships in the phrases from ones of logic to ones of time. Look at the example first:

**Logical**: given an API key, you can use the API.
**Temporal**: before you can use the API, you need to obtain an API key.
**Logical**: with successful user authentication, access to the system is granted.
**Temporal**:
**Logical**: upon completion of the testing phase, the software is deployed.
**Temporal**:
**Logical**: given that all tests pass, the feature branch is ready for integration.
**Temporal**:
**Logical**: with the resolution of all critical bugs, the software is ready for release.
**Temporal**:

**2b** Rewrite these sentences to make the relationship of cause and effect clearer. Use terms like *due to* or *because* to alter the relationships in the phrases from ones of logic to ones of cause. Look at the example first:

**Logical**: in the presence of memory leaks, the application experiences crashes.

**Causal**: the application crashes due to memory leaks.
**Logical**: with efficient code optimization, the software runs faster.
**Causal**:
**Logical**: in the absence of proper input validation, security vulnerabilities arise.
**Causal**:
**Logical**: with regular maintenance, the system's performance remains optimal.
**Causal**:
**Logical**: upon the detection of a critical bug, the deployment is halted.
**Causal**:
**Logical**: in the presence of outdated libraries, the software becomes vulnerable.
**Causal**:

## Works cited

1  N. M. Power, *A grounded theory of requirements documentation in the practice of software development* (Ph.D. dissertation), Dublin City University, 2022. [Online]. Available: https://doras.dcu.ie/18161/

2  C. Gralha, R. Pereira, M. Goulão, and J. Araujo, "On the impact of using different templates on creating and understanding user stories," in *2021 IEEE 29th International Requirements Engineering Conference (RE)*, Notre Dame, IN, USA, 2021, pp. 209–220. https://doi.org/10.1109/RE51729.2021.00026

3  W. Bhawna and B. Pawan, "Converting epics into user stories in Agile," *Global Sci-Tech*, vol. 11, no. 2, pp. 75–81, 2019. https://doi.org/10.5958/2455-7110.2019.00011.9

4  A. Boyarchuk, O. Pavlova, M. Bodna, and I. Lopatto, "Approach to the analysis of software requirements specification on its structure correctness," in *International Workshop on Intelligent Information Technologies and Systems of Information Security*, 2020, pp. 85–95.

5  K. Sawyer, "Mystery of orbiter crash solved," *The Washington Post*, 1 Oct. 1999. [Online]. Available: www.washingtonpost.com/wp-srv/national/longterm/space/stories/orbiter100199.htm

6  T. Øvad and L. B. Larsen, "The prevalence of UX design in agile development processes in industry," in *2015 Agile Conference*, National Harbor, MD, USA, 2015, pp. 40–49. https://doi.org/10.1109/Agile.2015.13

7  M. P. Robillard, "What makes APIs hard to learn? Answers from developers," *IEEE Software*, vol. 26, no. 6, pp. 27–34, 2009.

8  G. Uddin and M. P. Robillard, "How API documentation fails," *IEEE Software*, vol. 32, no. 4, pp. 68–75, 2015. https://doi.org/10.1109/MS.2014.80

9  J. Kuzmina, "Genre analysis of quality assurance (ISO 9000) documentation," *Baltic Journal of English Language, Literature and Culture*, vol. 3, pp. 76–86, 2013. https://doi.org/10.22364/BJELLC.03.2013.06

10 J. Nindel-Edwards and G. Steinke, "The development of a thorough test plan in the analysis phase leading to more successful software development projects," *Journal of International Technology and Information Management*, vol. 16, no. 1, pp. 65–72, 2007. https://doi.org/10.58729/1941-6679.1188

11 IEEE, *29119–3–2021 – IEEE/ISO/IEC international standard for software and systems engineering – software testing – part 3: Test documentation*. IEEE, 2021.

# Chapter 6

# User documents in SDLC

## 6.1 Introduction to user documentation

This section looks at the kinds of documents that are designed for the people who will be interacting with your product or service once it has completed the SDLC. There are two main categories: end users and system administrators. End users come from very diverse backgrounds and may or may not have technical knowledge or experience. System administrators, on the other hand, will be working with your product or service in a professional and technical sense. System administrators typically have some background knowledge of IT. The chapter will begin by focusing on different types of end-user documentation before exploring some of the main types of administrator documentation.

## 6.2 End-user documentation

This section looks at documents written for product users. End users are usually people without specific technical or professional knowledge of the product. They also often use the product in everyday professional or domestic circumstances. Think of the applications on a personal or mobile device, or the software used in a professional, non-IT setting. The uses of such applications can vary from personal health and finance to dual authentication applications for the workplace. Regardless of the use, your end user will need to know how to install, set up, and use this application.

### Importance of writing effective instructions

All end-user documentation is in some sense instructions. Effective instructions are important and are a key driver of end-user satisfaction [1]. They help people install and use new software applications and set up technical products. Imagine you are given a box of Lego to assemble, but you are not given the instructions. If the task is very, very simple, it might be easy to complete. If the task is even a little bit more complicated, however, it will be almost impossible to build the correct end result without carefully following the steps in a strict sequential order.

Now imagine that it is not a toy that you are constructing, but a system or piece of equipment that helps your organization to operate efficiently. Imagine that if you do not complete the task correctly, there is a risk of damage to the system and peripheral equipment, which carries knock-on effects of loss of productivity and possible data leakage.

Instruction manuals and other end-user documents are not only an opportunity to guide the end user in successful installation. They are also a way to explain the proper use and care of

DOI: 10.4324/9781032647524-6

your product and service as well as warn your end user against incorrect or improper care or usage. This can be especially useful to your organization in protecting against legal complaints that can be costly.

Instructions are important to get right. There are a lot of products available that can help you to write effective manuals, but it is a good idea to have the skills and knowledge for yourself. We begin by exploring some of the different types of instructions that can be found in end-user documentation for IT.

## Type of end-user documentation

Some examples of end-user documentation can be found in Table I.

## Writing for audience and purpose

The main user of a service or product is the intended audience of end-user documentation. Users are assumed to be general, which means using clear language and avoiding complex or technical terminology. Users may have some experience with operating a software application like the one you are introducing. However, they may also be completely new to the actions you are guiding them through. The rule of thumb, therefore, is to write for the second type of audience, without using condescending language. Remember that your audience may not be reading your documentation in their first language, so end user documentation should be written using simple and concise language. That helps everyone to understand and reduces any risk of alienating your readers.

### Purpose

End-user documents aim to provide clear directions for the use and care of technical products and services. They tell the user how to complete basic functions, how to troubleshoot any problems that occur, and provide extra information about functionality and/or customization. This practical purpose informs what is included, the style of language used, and how information is arranged. All of these aspects are carefully considered to maximize readability.

### Contexts of use

A key defining feature of instructions is that they are nearly always read while completing the task. Some audiences will begin with the instructions, reading everything before beginning, but they are extremely uncommon. Most readers will begin with step one and complete step one before reading the next step. Some readers will try to complete the process on their own before running into trouble and turning to your instructions. There are even some readers who will delegate the task to someone who is not the primary user. In many contexts, older relatives will ask younger relatives to help with the set-up and configuration of their personal devices. You need to write instructions that fit all possible audience types.

### Importance of design

As we have seen, your reader is likely to read a step and then complete the step. This stop-start nature means that your reader will lose a lot of time unless you design it to be jumped out of

Table 1 Types of end-user documentation

| | Audience | Purpose | Layout |
|---|---|---|---|
| **Quick-Start guide** | These are for new and inexperienced users of a product or service. | These are shorter versions of traditional manuals. They provide basic information that aims to help users quickly set up and start using a product or service. | They include headings, limited text, and procedures that are sequenced and use command voice. Graphics are a key part of quick-start guides. |
| **Instruction manual/ user guide** | These are for new and inexperienced users, and also for users who want to know about a less familiar feature or function. | These are comprehensive documents that provide step-by-step guidelines for how to use a product or service. | They include title pages, table of contents, headings, numbered steps, and frequent and signposted use of graphics. |
| **Workplace procedural document** | These are for professionals who need guidance in how to complete or follow a particular guideline at work, for example, requesting a benefit that is context specific. | These guidelines explain to employees how to complete a given task. They assume the reader already has some institutional knowledge about where to find resources. | They often include context-specific language and acronyms that are understood by members of the professional organization. Headings, numbered lists, and brief explanatory paragraphs written in general language are also common features. As procedural documents are typically online, they may also include hyperlinks that take the user to a specific site or related policy. |
| **Tutorial** | These are used by readers who value instructional content in additional formats. They may require content in a different format for accessibility reasons (e.g., users with a visual impairment or delayed processing when reading), or they may learn best by watching or hearing. | These are interactive guides that use multimodality to guide the user through a process like setting up an account or installing software. They use a combination of audiovisual and text content to make a process simpler for the user. | These usually involve video of a live individual, a screen recording, or animation that walks the user through the process they are trying to complete. Longer tutorials include bookmarked timestamps that help the user to skip forward to relevant content. |
| **Online support** | People using online and technical products for their work, study, or personal uses. | This is usually context-specific support that is embedded within the design of software applications themselves. It provides just-in-time support for using a feature or completing a process. | Online support typically uses searchable lists that are topically organized and listed in alphabetical order. Sometimes examples are included to clarify. General language is used to promote accessibility. |

and into easily [2]. There are some key design components that can help make it easier for your audience to quickly find the next step without losing time. More detail about the various design principles was covered in Chapter 2, but instructions are a place where the principles of white space, alignment, consistency, and contrast are all very important.

## Instruction manuals and quick-start guides

The first category of end-user documentation includes instruction manuals and quick-start guides. Instruction manuals are also often called user manuals and tend to be more complete documents that guide the appropriate and safe usage of the product or service. Quick-start guides (QSGs) are typically much briefer and are designed to help the user quickly set up the device and begin using it. Regardless of the name, these documents help the end user to set up and use their product or service.

As we have already seen in Chapter 1, there are many types of audience in IT contexts. Unless the product is extremely specialized, the end user for a piece of software or a technical product is likely to be varied. Best practice is always to write for the most general audience in order to avoid unintentionally alienating some users.

### Organizing and formatting instruction manuals

Instruction manuals and user manuals tend to follow this structure:

1 Cover Page
2 Table of Contents
3 Product Overview and Introduction
4 Equipment/Tools/Safety/Software Requirements
5 Installation Guide
6 Account Set-up Guide (optional)
7 Basic Usage Guide(s)
8 Additional Usage Guide(s) (optional)
9 Troubleshooting and Extra Help

Each of these is discussed in more detail as follows:

**1 Cover page**
The cover page does not need a lot of detail, but it should clearly identify the product or service it is created for and the purpose of the document. The cover page may identify which version of the software or product it has been created for. Sometimes the operating system it is associated with is highlighted, and some instruction manuals also designate a specific reader, usually when the programme or product is more technical and not intended for general use.

Remember to incorporate the design principles into your cover page. It should clearly indicate the document's purpose – informative titles are more useful to your reader than generic titles. Most manuals include branding such as logos or the app's icon or an image of the product. Colours are used to add unity and connect the product, software, or service to your brand. Your company or organization's name is also usually visible.

## 2  Table of Contents

While it may seem obvious, each user manual should begin with a table of contents. Remember that not all of your users will be starting at the beginning of the task. For those users who need to begin reading the instructions midway through a task, the table of contents will help them find what they are looking for without needing to flip or scroll through the whole document. Some users will not require help with installation, or they may already have an account. Some users will only want to find out how to use an advanced feature or will be looking for where to find additional help. The table of contents facilitates better navigation for all users.

The table of contents should be clearly organized and formatted using design principles such as consistency and alignment. Consistency helps your reader to read vertically through the manual and alignment helps them to understand how the parts relate to one another. Don't forget to include page numbers – a table of contents without page numbers is useless for your reader!

## 3  Product overview and introduction

Instruction manuals and quick-start guides begin with a brief overview of the product or service and its use. The overview should include any hardware or software your user needs to have installed to be able to run the product or service. It lets your user know whether their current device or system is compatible with the product or service – if your application is only compatible with the Linux OS, for example, it should be clearly stated in the product overview. Remember: this should not be extensive, and it should be written with your audience in mind – a general audience. The purpose of the overview and introduction is not to provide an exhaustive list of all the products' features and capabilities, but to let your end user know what it is designed for and how they can use it in the personal, professional, or academic lives. Quick-start guides tend to be even briefer and may limit this to one to two sentences.

This information should be presented in narrative paragraphs. You should use an informal tone, appropriate for a general reader. You language (see Chapter 3, language focus) and contractions are acceptable when discussing your product or service.

If you want to include a list of features, functionalities, or specifications, these can be grouped logically and bulleted to facilitate easier vertical reading.

## 4  Equipment/tools/safety/software requirements

Before your user begins the process, it is very important that they know what they need to have on hand in order to be successful. If they need to configure hardware, list any tools or equipment they will need. Sometimes specific conditions are required, like a long table, cool temperature, well-lit space, or good ventilation. Some products require special safety information, like equipment that could pose severe shock risks or risks of electrocution or other bodily harm if not handled carefully. Safety information should be clearly outlined before your user begins the installation task.

When it is a software program that your user is installing, physical dangers are less important, but some conditions may still need to be met for success. If your user needs a particular version of operating system or if your product is only compatible with specific operating systems, this should be reiterated here. While physical safety is not usually a concern, maintaining personal data security is very important any time a user's personally identifiable details are stored. Data security practices and strong password guidelines are frequently included here.

Whether it is the physical safety, conditions, or online systems and security that you focus on, your user needs to be aware of this information before they begin to install, set up, or use the product or service. That is why this information should come before directions for installing or using the product or service.

Information in this section should provide clear indicators of the dangers associated with not following the guidelines. Lists of required equipment, tools, and/or conditions should be bulleted for easier scanning. Important safety and security information should be emphasized using a number of highlighting features. Warnings and cautions should be clearly labelled.

## 5  Installation guide

Once your user knows what the product or service is for, the system requirements needed, and is aware of any pertinent guidelines, they are ready to install the programme or set up the hardware. Installation should include all steps from start to finish, without omitting any obvious steps.

The section should begin with a content-oriented heading that clearly indicates what the user will accomplish (see Chapter 2, Section 2.3). The heading should correspond to what is used in the table of contents. Next, some manuals also include a sentence that clearly states what will be achieved. While this is optional, it is another way to make sure your manual is clear and reader-friendly.

Number each step in strict order and ensure that there is only one action per number. Steps should be written using command voice or the imperative (see Language Focus later in this chapter). If there is additional information, for example, guidelines or suggestions on how to configure settings, these should be grouped near the step, but they should not be numbered. They should be labelled as guidelines or notes.

Illustrations that are used to provide a visual cue should be grouped near their associated steps. Some manuals also point to the image, by referring to its figure number in brackets at the end of the step. Remember your design principles here – images should never be placed on the left with a document written in English. Since they attract the eye more than text, they should be placed to the right of text or below it.

Steps should read vertically down the page, so skip a line after a step and start with the next step on a new line. Some manuals use line breaks to mark the beginning and end of sections.

## 6  Account set-up guide (optional)

The next set of instructions to give your user includes information about how to set up any associated accounts. Apple, for instance, guides users through how to create an Apple ID as the next major part of setting up any of their devices. It is important to include data security guidelines and any password or username conditions on an as-needed basis, so be sure to include them here.

This section will follow a similar format as the installation section. Ensure the heading matches what is presented in your table of contents and be consistent with design of your headings. Follow the same guidelines for creating the steps associated with account creation, and if you include any visuals, make sure they are also appropriately grouped and placed on the page. Repetition of formatting is key here to ensuring your instructions demonstrate readability (see Chapter 2, Section 2.2).

## 7  Basic usage guide(s)

Now that your user has installed the software and created an account, they are ready to use the programme or product. The next section will depend on your product, but the primary use of the application or product should be explained using the same careful sequencing and formatting (explained later). This might include the features and functionalities associated with the product or software or service. Basic uses should be subdivided from one another and clearly signposted. Each use should be included in the table of contents at the beginning of the manual, so they are easy for your user to find.

Once again, this section will be formatted very similarly to installation and account set-up sections. Headings, numbering, sequencing, and design will also be similar. Don't forget to limit one step to each number and to use appropriate grammar. Make sure visual cues are placed where they can best help your reader complete the task.

If your product or service is a platform with various functionalities, explain the functions and where/how to find them. If there are any special navigation strategies, include those in this section.

## 8   Additional usage guide(s)

More comprehensive instruction manuals will include a complete list of the functions and features of a product or service. Secondary usage guidelines should come after the primary usage information. Some logical ordering should be used, whether sequential, spatial, or chronological.

These are highly context-dependent, based on the purpose and use of your product or service. As before, if there is a procedure your reader needs to complete, provide it using the same format used in installation and use. Use headings and alignment to help your reader connect ideas to one another. Provide logical and clear guidelines and ensure the additional usages are listed in order in the table of contents.

## 9   Troubleshooting and extra help

Most instruction manuals include a section providing troubleshooting advice and where to find additional support such as toll-free numbers and online support. Such information can be embedded within the document or can be linked to at an online or digital source. This section typically comes at the end of the manual and should be structured so that it is easy to scan [3]. Whatever organizational scheme is used to order troubleshooting areas (such as topical), the areas should be listed alphabetically so that users can easily find what they are looking for.

> **Key vocabulary**
>
> Sequential – in order of process, e.g., first, second, third . . .
> Spatial – according to how information/components are organized on the page/screen, e.g. top left, bottom right . . .
> *Chronological – according to time-order, e.g., first, after 10 minutes . . .*

### Effective and ineffective layout

A good example of an everyday procedure is following the cooking instructions in a recipe. Recipes usually demonstrate effective layout, although some sources do not take the reader into account, offering instructions in a paragraph format that is harder to read while doing.

Compare the following two examples:

## Example 1: Classic pomodoro sauce recipe

---

To begin, you will need to gather 900 grams of fresh tomatoes. You could also use a can (28 ounces) of whole peeled tomatoes. You will also need three to four minced garlic cloves, a quarter cup of extra-virgin olive oil, one teaspoon each of dried oregano and basil, and around half a teaspoon of red pepper flakes (these can be adjusted according to your preference). You should also have salt and pepper ready to add for taste. Some people also include a quarter cup of chopped fresh basil leaves, while others opt for a quarter cup of grated Parmesan sauce.

---

Here, everything that is needed is given, but it is given in a paragraph. This means your reader will need to read across the paragraph, and probably read from beginning to end to make sure they have all the ingredients and equipment they need. For example, your reader does not know that you may add fresh basil leaves or grated Parmesan until the last sentence.

## Example 2: Classic pomodoro sauce recipe

---

### Ingredients

- 900 grams of fresh, ripe tomatoes or one can (28 oz) whole peeled tomatoes
- 3–4 garlic cloves, minced
- ¼ cup extra-virgin olive oil
- 1 tsp dried oregano
- 1 tsp dried basil
- ½ tsp red pepper flakes (adjust to taste)
- salt and freshly ground black pepper, to taste
- OPTIONAL: ¼ cup chopped basil leaves
- OPTIONAL: ¼ cup grated Parmesan cheese

Cooking instructions

1  Peel and chop 900 grams of fresh tomatoes (or use a 28 oz can of whole peeled tomatoes) and place on the stovetop in a medium-sized saucepan over medium heat.
2  Mince 3–4 garlic cloves.
3  Add garlic, olive oil and herbs to the tomatoes.
4  Season to taste with salt and pepper.
5  Bring to a boil and reduce heat.
6  Simmer for 15–20 minutes.
7  Remove from heat and serve immediately atop fresh cooked pasta or let cool and transfer to storage containers.
8  Optional: add ¼ cup chopped fresh basil leaves or ¼ cup of grated Parmesan cheese.

---

The second example is more reader-friendly. Your reader can read vertically as opposed to horizontally. It provides a list of ingredients and quantities ahead of the cooking instructions in an easily scannable list [4]. This means your user is less likely to begin the process without knowing which ingredients they need to have available. Alternative ingredients are clearly labelled as such. Steps are presented in a strict sequential order, with one action per step. Headings have been used to distinguish the ingredient list from the procedure, and optional steps have been clearly labelled. This is a better example of providing a procedure in a reading-while-doing format.

*Additional content in instruction manuals*

Some manuals include additional information. This is context and product/service dependent, but some possible additional sections and their purpose are listed as follows:

- **Glossary of terms** – if the product or service requires the use of specialized terminology or jargon, these are commonly explained in a separate section that users can refer to. This information is listed and organized alphabetically, making it easier for the reader to find what they are looking for.
- **UI guidelines** – sometimes a user manual will include information about the product or service's user interface (UI), such as the layout, design, and basic navigation principles of a software product's user interface. Think about the basic actions your user may need to execute in order to move through your product. These should be explained and ideally examples should be given. Often, examples include basic motion and screen overlay to add non-textual cues.
- **Security and privacy documentation** – these sections provide your users with important information surrounding the product or service's data security features, how data is handed, and what privacy measures are in place. Make sure that this information is in keeping with the local laws of your end user. The European Union, for example, has strict regulations controlling what is and isn't allowed regarding cookies, storing, using, and passing along your user's personal data.

*Instruction manuals vs quick-start guides*

Remember, instruction manuals and user manuals are comprehensive documents. Quick-start guides are not. They are designed to get users set up and ready to use the basic functions of the product or service. Quick-start guides will not usually include detailed security information, guidelines on additional uses or functionality. Even more than other forms of technical communication, QSGs are designed to be economical and should be limited to only what the reader absolutely needs to know to get started. If it is not absolutely necessary to the basic set up and operation of your product or service, don't include it in your QSG!

*Organizing and formatting quick-start guides*

As we have already established, QSGs are brief documents – they may be a mini booklet or a folded paper packaged with a product. Information will be organized using headings and numbers that sequence steps and separate processes and features from one another. The specific format that you choose is less important than the fact that the QSG is succinct. While visuals are very important in manuals, they are even more important for quick-start guides. Images should be the main content and organizing feature of a quick-start guide. It might be helpful to think of it this way: whereas instruction manuals are text with visuals, QSGs are visuals with text.

If there is a process you are guiding your reader through, a flowchart or diagram structure can help to facilitate this in a visual format. If you are explaining the functionality and features, a labelled diagram with a key is an effective way of clearly demonstrating the various parts. Other organizational features that have been discussed are also key – things like using headings, pointing to the text, and grouping images and text together. The design principles are key to making an effective QSG that flows coherently, leading your user through the process or functionalities of your product or service. Technical writers often work in collaboration with technical illustrators to create these.

Good instructions checklist:
Is your end-user document . . .

- Designed for context and audience?
- Ordered appropriately in terms of content?
- Organized with action-oriented headings?
- Clearly sequenced using separate numbers for each user action?
- Organized using vertical listing?
- Full of white space to allow your users to do the task while reading?
- Written using clear and simple language use?

## Policy documents and standard operating procedures

A second category of end-user documentation in IT relates to policy documents and standard operating procedures (SOPs). Unlike instruction manuals and quick-start guides, these documents are not written for a general audience and their contexts of use is professional. Furthermore, both documents regulate acceptable work practices and therefore can carry more significant consequences, both legal and otherwise. We see this in the level of formality present in both document types – both types use a formal register (for more on formality, see Chapter 1, Section 1.5). Employees interact with these documents to perform their jobs correctly. However, that does not mean they are always written for an audience with a strong background in IT.

### Writing effective policy documents

Like other technical documentation in IT contexts, policy documents should be written in a reader-friendly style that can be understood by all members of the organization, regardless of their IT background [5]. This means using plain language and providing definitions of key terms whenever possible.

IT policy documents should provide clear guidelines of what constitutes acceptable and/or unacceptable use. There should be no room for interpretation or misunderstanding of what is and is not permitted. However, the policies should not be so restrictive that they quickly become redundant or do not reflect the changing nature of IT in the workplace. Because of the ubiquity of technology in all forms of workplace, the types of policy document that may exist are equally varied. A partial list is outlined in what follows:

- Acceptable Use
- Bring Your Own Device (BYOD)
- Change Management
- Cloud Computing
- Data Retention
- Disaster Recovery and Business Continuity
- Incident Response Plan
- IT Asset management
- IT Emergency Response
- IT Employment
- IT Security

- IT Software management
- Network Security
- Password
- Privacy
- Remote Access
- Social Media
- Software Development
- Vendor and Third-Party Security

### Organizing and formatting policy documents

Policy documents are generally relatively brief, limited to one to three pages. The procedures that are often associated with them can be much longer. Policies operate at a general level, so they do not include specifics and details. You and members of your organization can expect to find those in the procedure documents that relate to each policy.

Policy documents will be formatted according to the specific conventions of your organization and to the needs of their content. However, they commonly include the following eight sections:

1 Title (not cover page)
2 Record of amendments and revisions
3 Purpose
4 Scope
5 Policy
6 Related Policies and Laws (as applicable)
7 Directly Responsible Individual (DRI)
8 Revision History

### 1 Title

The title of your policy will follow your organization's conventions for policy documentation. Many organizations include identifiers that help to classify what type of policy it is, but this is optional. What is important is that your title is informative.

Examples:
IT Policy IT5601 Acceptable Usage
Policy Acceptable Usage
Acceptable Usage Policy

### 2 Records

This is often formatted like the top of a memo where details of the sender, receiver, and date are organized. Sometimes it is presented as a table. This part of the policy document provides administrative and organizational information and identifies who is responsible for creating the policy, who owns it, and relevant dates, including creation and most recent revisions. See Table II for an example.

### 3 Purpose

This section provides a very brief overview of the reason for the policy. This is provided in narrative form. It is written in one to two sentences. It should explain the reason the policy

*Table II* Example of records

*Acme Corp*

| Category | IT assessment management | Policy number | ACP-ITS-041 |
|---|---|---|---|
| Distribution | Internal | Version | 1.2 |
| Responsible | CTO | Policy owner | IT department |
| Approval Date | 23 February 2020 | Effective date | 23 February 2020 |
| Last Reviewed Date | 23 February 2023 | Next review date | 23 February 2026 |

exists and how it relates to your organization's operations. Language should be precise and signposting should be used to clearly convey the purpose.

Example:

The purpose of this policy is to outline the acceptable use of all information technology assets at Acme Corp. It also serves to provide directives and guidelines towards all usage of services and systems pertaining to the organization's IT infrastructure.

## 4  Scope

This section should identify who the policy applies to and which IT assets or systems are covered by it. If relevant, it will also identify who the policy applies to. The scope is another brief section. Again, clear language indicating the scope is a valued means of making the document reader-friendly.

Example:

This policy applies to all Acme Corp information technology assets, systems, networks, and environments. It applies to all users of said assets, systems, networks, and environments, whether full-time or contractual employees.

## 5  Policy

This section depends largely on the subject matter. Simple, umbrella policies may just be listed numerically. More complex policies might be organized topically and then delineated numerically. Specific obligations and prohibitions should be outlined in a bulleted list with the associated policy point. The following example demonstrates how a more complex policy includes topical organization. Note that formatting issues such as sequential nested listing and alignment have been used to help the reader understand how the parts of the policy work together.

Example:

*3.1  General use and ownership*

  3.1.1 Acme Corp's IT assets are intended for legitimate business use.
  3.1.2 Personal use is permitted only to the extent where such use does not interfere with an individual's professional performance or obligations, does not restrict Acme Corp's operations, does not pose any risk to Acme Corp, and does not breach Acme Corp's internal policies or legal obligations.

3.2 *General security requirements*

    3.2.1 Any and all software use must be provided and licensed by Acme Corp. All individuals must only use software that has been approved by Acme Corp's IT department.

    3.2.2 Employees are prohibited from downloading web-based applications without the express written consent of the relevant IT officer.

    3.2.3 Users will maintain up-to-date antivirus protection as provided by Acme Corp.

    3.2.4 Users will refrain from storing sensitive and/or confidential data on Acme Corp devices without proper encryption and passwords in place.

3.3 Specified details of the areas subject to the category of acceptable use are outlined in the associated procedures.

3.4 Employees and/or users found to be in violation of this policy will be subject to disciplinary action, including the possibility of suspension and/or termination.

## 6 Related policies and laws (as applicable)

This section refers your employees to any relevant policies within your organization. They may be associated IT policies, or they may be policies that come from other parts of your organization, like HR. The aforementioned policy, for instance, mentions passwords and legal obligations. If there are policies related to these areas, you should direct your reader to read them or at least be aware of their existence. The related policies and/or laws should be listed using their official names and any identifiers.

Example:
Related Policies and Laws

– ACP-HR-103
– ACP-ITS-032
– Telecommunications Regulatory Act and Amendments (2018)

## 7 Directly responsible individual (DRI)

In some organizations, a specific individual holds the responsibility for managing and implementing a policy. More frequently, it is the responsibility of an office or a department. Sometimes this section of a policy is labelled 'Administration'. Crucially, the entity is accountable for managing the policy, its distribution, and review.

Example:
This policy is administered by the IT department. Any queries can be directed to the IT Manager or the Chief Technology Officer.

## 8 Revision history

This section is usually formatted as a table that indicates review dates and lists any associated changes. If there are multiple changes or the revision has been particularly complex, they may be topically organized. If no major changes have been made, that will be noted. Note that revisions will be listed from the most recent to the oldest. See Table III for an example.

### Policy documents and SOPs

Frequently, workplaces include a sort of hybrid policy document called a standard operating procedure. These are policy documents that outline how complex, high-stakes procedures are to be conducted. They ensure quality, consistency, and reduction of error. SOPs contain much

*Table III* Example of revision history

| Date | Revision |
|---|---|
| **23 February 2023** | Minor revisions and non-substantive changes approved by IT Manager and Chief Technology Officer. |
| **23 February 2020** | New policy approved by Chief Technology Officer. |

of the information of a policy document, but they also include research and reference to other quality control standards. In this chapter, SOPs will be covered with the second major category of user documents: system administrator documentation.

## FAQs and self-access IT resources

Increasingly, user documentation for IT contexts has become self-access and digitized. This content is usually hosted on your organization's website, often nested within pages connected to your IT department. This could be publicly available on the world wide web or restricted to internal use via an intranet. Regardless, such materials help your users to troubleshoot IT problems, helping your organization to run more smoothly and building trust in your content. These benefits only happen if your users are able to find your FAQ page. So make sure this page is easy for them to find, easy to use, and you will reserve the efforts and skills of your technicians and support staff for more complex problems. Typically, users expect to find the FAQ page via your website menu or in the footer at the bottom of your webpage.

### Writing effective FAQs and self-access resources

The content of your FAQ page will depend on your business and your clients/users; however, there are a few suggestions for how to decide what to include:

- Look at the questions your clients/users often ask
- Look at the FAQ page of your competitors' websites
- Ask your IT support department for the problems users most frequently encounter
- Include multiple synonyms to maximize the likelihood of users finding what they need

It is also a good idea to include a search bar where users can type in their own queries. This will be more user-friendly if you have included multiple synonyms for each of the topics in your FAQ or self-access section.

### Formatting your FAQs and self-access resources

Your page should be organized using many of the general guidelines for effective procedural writing. Alphabetical listing, effective use of headings, and good alignment and consistency will make your FAQ page easier to navigate. Any series of steps your user needs to complete should be formatted according to the guidelines of an instruction manual or a quick-start guide, depending on purpose and context. Including visuals like diagrams or screenshots will provide a cue that makes the text-based information easier to digest and more accessible for all readers.

Remember to incorporate the design elements of alignment, balance, and white space to make the page reader-friendly (see Chapter 2 for more discussion of these elements).

## FAQs and multimodality

Increasingly, self-access resources make use of multimodal communication or forms of communicating that use a combination of text, visuals, audio, and/or video elements to communicate (see Chapter 9 for a more comprehensive discussion of multimodal communication in IT). Video tutorials are a common self-access resource for many organizations. These can be highly professional or amateur, externally created, or internally produced. Typically, a video tutorial in a self-access resource includes voice-over narrative with a screen capture that helps your user to experience how to accomplish the desired task while being given a verbal explanation of what to do.

## Some important guidelines for procedural writing

Different types of end user documentation are differently structured, but they all share several common organizational features. While they have all been discussed previously, it is important to make effective use of each of these features in your end-user documentation. To ensure that your process is clearly structured, your procedure (the information explaining the steps the reader will follow) should follow the following guidelines.

### Guideline 1: Use headings

Headings have already been mentioned frequently. That's because they allow your reader to quickly find information that is relevant to what they want to achieve. Your users are motivated to accomplish a task, so you need to incorporate plenty of headings to help your readers to move more easily through the text to complete their task. Headings assist your reader with navigating into and out of the instructions while they are completing the task at the same time.

Effective headings are action-oriented because they focus on what your reader wants to do, making them more reader-friendly. Make sure your headings use consistent formatting from one section to the next.

### Guideline 2: Carefully sequence actions

Actions should be presented in a clear sequence. Begin at the beginning and do not omit or assume that the reader knows what to do next. If they need to click 'next', the sequence should tell them to do so. Numbering your steps is the clearest way to show your reader the order that steps should be taken. Each action should have one number.

Remember to use vertical organization by stacking steps on top of one another. This allows for the kind of vertical reading that is more efficient and easier to do while multitasking. Don't forget to also include white space that can help break up text and facilitate more efficient scanning.

Some end-user documentation will require carefully sequenced lists. Policy documents and SOPs, for instance, use listing to organize the major sections of the document as well as subsections. Sequence lists using numbers, letters, and/or Roman numerals is acceptable, although keep in mind that Roman numerals are not always as universally understood as numbering. Lists should begin with a number rather than a bullet point when the order matters and bullet points when the order is not important.

*Guideline 3: Use white space*

Actions the reader needs to take should be presented vertically, since reading across a paragraph requires more attention and effort than reading down a page. Use white space to help your reader find the next step easily without unnecessary repetition or time wasted finding the next step. If you have any associated information, it should be grouped together with the step.

*Guideline 4: Use visual cues*

Remember that not all of your end users will have the same understanding and confidence with the task. Providing visual illustrations is a great way of making your instructions more accessible. Use high-quality graphics. Any graphics you use should be grouped close to the steps to demonstrate by proximity that they are connected.

*Guideline 5: Be explicit*

Avoid having your reader guess. If they need to be aware of important safety or security information at a given time, say so in your written instructions. If they have finished the task, let them know with a clear sentence telling them there is nothing left to do. Be explicit. By clearly outlining what is necessary at the important points and when the process is finished, you make the sequence of steps easier to follow and more accessible for all readers.

*Guideline 6: Emphasize key information*

Often, end-user documentation includes information that can affect legal compliance, security, or data privacy. In these cases, it is important that the user does not miss this important content. When you have something especially important for your reader to notice, use emphatic features like bold, colour, and size to highlight the content and attract the reader's attention. The more important the content, the more emphatic the strategies used to draw your reader's eye!

Refer to Chapter 2, (especially the sections on contrast, emphasis, and movement) for further explanations and details of how these can impact how a technical document is read.

Following all of these guidelines should help ensure that your documentation is of good *functional* and *structural* quality. Functional quality is about whether a document achieves its goal or not, while structural quality is about whether it is clearly written [6].

## 6.3    Online tools

Many resources exist online to facilitate effective creation of user manuals and other forms of end-user documentation. Most of these are paid services that will provide limited use templates or demos. They incorporate the considerations in this book in an easy-to-fill template, offering the choice to modify design and create branded materials that are suitable for your purpose or workplace. They are a great place to start with drafting end-user documentation, especially for brainstorming ideas or modelling the type of language your user might expect. However, programmes like these also often contain major drawbacks that limit the functionality, design, and reader appropriateness.

## 6.4   Language focus for user documents

The following section provides useful language that is commonly found in user documentation. In end-user documentation where your reader is reading while doing, there are a few commonly used structures to be aware of:

- Command/imperative voice
- Sequential language
- Discussing the result of an action: two choices

  - By + gerund phrases
  - Infinitive phrases

Each of these are explained and illustrated in more detail as follows:

### Command/imperative voice

The first structure to be aware of is the command voice when telling your reader what steps to take. Avoid using *you* [7] and begin any actions the user must take with an imperative verb. Table IV shows the difference.

Using command voice keeps the focus on the action the reader needs to take.

### Sequencing language

The second language point that is especially important in end-user documentation is the use of sequencing language. Use sequencers to show the order that should occur when two actions are related.

**Once** the installation is complete, remove the installation package.
**After** the programme has updated, restart the computer.

Note that sequencing language combines with the imperative in the previous examples to relate a user action within the context of an already in-progress task.

### Discussing the results of an action: two grammatical choices

A common component of any procedure is to explain to your reader what will happen if or when they follow your directions. To do this, we express the result of an action. There are two structures that are commonly used in technical communication: gerund phrases and infinitive phrases. Table V illustrates how and when they are used.

For more discussion of information flow, see Chapter 7, 8.

*Table IV* Command vs indicative mood

| Command voice | Indicative voice |
| --- | --- |
| *Open* the browser | You need to open the browser. |
| *Accept* the terms | You must accept the terms. |
| *Select* your department | You should select your department. |

*Table V* Gerund and infinitive phrases

|  | Gerund Phrase | Infinitive Phrase |
|---|---|---|
| **When to use** | Use a gerund phrase when you want to emphasize the work or effort involved in complete a task. | When you want to emphasize the end goal or objective of a task, use an infinitive phrase. |
| **How to create** | Add -ing to By +verb | Use To + verb phrase, provide the task required. |
| **Example** | *By spending hours and hours studying and practising, you, too, can learn LaTeX.* | *To make sure your website provides a good user experience, be sure to consider your API.* |
| *Example in reverse* | *You, too, can learn LaTeX by spending hours and hours studying and practising.* | *Be sure to consider your API to make sure your website provides a good user experience.* |

Note: both of these structures can be reversed by switching the order of the clauses and removing the comma.

## 6.5   System admin documentation

The second major category of user documents in the software design life cycle involves system administrator documentation. The main difference regarding these documents is audience. System admin documentation is written for an audience with IT knowledge and experience. System administrators work in IT, so their knowledge and contexts of use will be different from a professional or general user of a software program. They assist in the successful operation of IT-related systems like your organization's networks and security systems. Despite their expertise, however, they, like everyone else, require good documentation to do their job efficiently and effectively [8].

While a system administrator's duties can and do vary from organization to organization, they typically oversee the installation, upkeep, and renewal of software packages used by your organization.

### Common types of system admin documentation

There is a wide range of sysadmin documentation, depending on the duties of your system administrator. Table VI provides several common types, their definitions, and some examples.

### Access control privilege documentation

In the previous table, access control documentation is identified as very important to maintaining the security of your organization's IT infrastructure. This involves limiting who can access what and when or how. Access controls can be based on different models, and they can limit access to physical contexts, to times, or to roles.

For example, universities sometimes limit online tests to the IP address of the physical campus so that students can only access the test when they are on site. This is an example of limiting physical access. The same test may only be available during a specified time period, which

Table VI Types of system admin documentation

| Document type | Definition | Example |
|---|---|---|
| Installation guides and configuration manuals | Installation guides and configuration manuals are manuals and instruction guides for system administrators to oversee how to configure and install the systems and software used by an organization. They differ from installation guides for general users in terms of complexity and comprehensiveness, with each section including subsections that may span several pages. | The system admin guide for an operating system includes sections detailing:<br>– Basic configuration settings<br>– Subscription and support<br>– Installing and managing software<br>– Infrastructure services<br>– Servers<br>– Monitoring and automation<br>– System backup and recovery |
| Performance tuning and trouble-shooting guides | These guides are concerned with the performance of your system. They provide sysadmin with strategies to optimize the operation of computing systems, networks, and applications. They help sysadmin to identify bottlenecks and usage inefficiencies and find ways to maximize resource utilization. | Performance tuning and troubleshooting guides may help sysadmin to identify CPU optimization goals, providing coding snippets to modify and finetune specific settings, balancing the considerations of performance, security, and stability. |
| User access and privilege documentation | Another key category of sysadmin documents involves access control policies or documents. They outline what constitutes acceptable use and what limitations and/or controls are placed on individual users, according to their role within your organization. The aim of this is to protect your organization's IT infrastructure and data. | One example of user access documentation is called role-based access control (RBAC). In this model, users are assigned to different groups, based on specific levels of permissions. See what follows for more discussion. |
| System architecture documents and network topology maps | These two types of documentation provide sysadmin with a snapshot of how a system is structured. Both types provide a blueprint or map of the hardware of your organization. System architecture documentation is a broader picture that includes both software and hardware that includes both software and hardware along with their interactions. Network topology maps offer a more specific focus on the network layout, illustrating the interconnectivity of the various network devices. | A basic example of a network topology map for a home wi-fi router demonstrates the areas covered by this type of document (see Figure 6.1). In this example, we see the wi-fi router providing internet connectivity to household devices.<br>A system architecture diagram of the same home wi-fi would include the same as the topology map, but also include specifics relating to the modem – its antennae, processing power, etc. It may include overlay boxes to show software requirements and OS for various devices. |

Table VI (Continued)

| Document type | Definition | Example |
|---|---|---|
| Security policies and procedures; compliance | These documents are similar in purpose to those written for a general audience, however, security policies and procedures that are written for sysadmin are typically more technical and complex. They include details related to network configurations, access roles, and handling administrative tools. Their language and use of technical terminology illustrate their intended audience and use case. | Security policies and procedures for sysadmin users typically<br>• Outline roles and responsibilities.<br>• Describe specific data maintenance and security protocols, including incident response.<br>• Comply with external regulatory authorities according to specific industry (e.g., healthcare, banking).<br>• Outline the certifications that sysadmin users should maintain current in order to stay up-to-date in their practice. |
| Backup and recovery plans | An important category of sysadmin documentation includes backup and recovery plans. The purpose of these documents is to provide contingency plans and practical steps that should be taken by the system administrator in case of system disruptions, user errors, or security breaches.<br>Backup and recovery plans should outline where and how data copies can be accessed to restore operations. | Effective backup and recovery plans have been designed to prevent data loss that could disrupt how your organization runs.<br>A well-known example is the 3-2-1 method: you should have three copies of your data, using two different media, with one off-site copy for disaster recovery. Technological advances have revised the recommendation, for example, suggesting you should keep an additional cloud-based copy of data in a separate geographic location. |
| Change management procedures | Another type of document includes change management procedures. Sysadmin use these to ensure that infrastructure changes happen systematically and without unforeseen errors. To achieve this, they have to consider the impact of the proposed change and conduct a cost-benefit analysis as well as highlight any potential risks. Depending on the size and scope of your business, these can be general or very detailed guidelines. This is determined by how costly any downtime could be in terms of money and/or effort. | Many organizations outsource the tracking and workflow connected to IT change management procedures. A common example is to use JIRA to track and document such changes. Sometimes this is used together with another knowledge management platform, such as Confluence. Other organizations have word-processing documentation that details changes and actions taken, saved in a fileshare location. |

(Continued)

*Table VI* (Continued)

| Document type | Definition | Example |
|---|---|---|
| Service level agreements (SLAs) | These documents outline the agreed expectations of an external or internal provider of a system or service. Think of them as IT-focused contracts that provide clear points about the service to be provided, how it will be measured, and accountability for any failures to provide the service. | Google provides an extensive list of SLAs online. An example is the SLA for SQL, which explains the terms of the service for the client, the financial credits that a client will be eligible for if the terms are not upheld on Google's end, and also provides definitions of key terms. |
| Audit trails and logs | Audit trails and logs are important records that system administrators can access to find all system activity. This is especially important in identifying breaches, significant user errors, and unauthorized access. The audit trail log provides a digital footprint of all user activity. | Audit trail logs should detail application activities including:<br>– Login and log-off activity (successful and unsuccessful attempts)<br>– User-initiated actions<br>– Data accessed (successful and unsuccessful attempts)<br>– Configuration changes (successful and unsuccessful attempts). |

would be an example of time-based limits. Finally, the test would only be available to students enrolled on the course, which is an example of role-based access.

In role-based access control (RBAC), everyone employed or affiliated with your organization will be assigned a category and associated permissions and constraints will be placed on what they can and cannot do on your system.

Users are frequently assigned roles such as

- **Administrator** (full system access, including modification of user permissions, management of security settings)
- **Manager** (access to department resources, editing capability for department data and system configuration)
- **Staff** (access to specified applications necessary for job functions)
- **Guest** (limited access to non-sensitive data and applications, no modifiable capabilities)

### Network topology map example

Table VI also explains how sysadmin use network topology maps to illustrate hardware connectivity within an organization. Figure 6.1 provides an illustration of the associated example in Table VI.

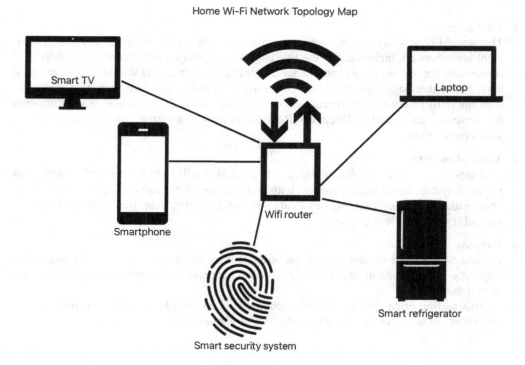

Figure 6.1 Home wi-fi network topology map

*Writing effective SOPs*

The purpose of a standard operating procedure (SOP) document is essentially quality control. The aim is to provide the most effective, efficient way of completing workplace tasks so that they can be done to the highest level possible each time. SOPs also play a role in compliance and quality control. They are complex guidelines for how to perform complex procedures in the workplace. While there are many web-based services that offer paid templates for creating industry-specific SOPs, it is best to write your own that is appropriate to your context, purpose, and audience.

SOPs are research-based, so they are evidence-based and hold strong credibility within their workplace. Because they provide guidelines for how best to accomplish a workplace task, it is important that your SOP is up-to-date. Regular revisions should be scheduled. Any revisions should be clearly noted in the document review section. SOPs exist in almost any professional setting, and they are important technical documents in a variety of applied IT contexts.

The most basic structure for an SOP includes the following six components:

1  Title Page
2  Table of Contents
3  Purpose
4  Procedures
5  Quality Assurance/Quality Control
6  References

## 1  Title page

This should clearly identify the area and procedure that are the focus of this SOP. Document identifiers are included, and the title page should also include the office or individual responsible for overseeing its compliance. Any necessary approvals should also be included here. Some title pages include document revision records. These are formatted as a table with the most recent to least recent versions ordered. Some SOPs include this information in a separate section called 'Document Revision History'. Be sure to use the conventions of your organization.

## 2  Table of contents

Like many other longer IT documents, an effective SOP will include a table of contents that provides system administrator readers with an overview of the entire document. This helps your reader to find relevant sections easily and avoid wasting time. Remember: this is only helpful if you provide page numbers!

## 3  Purpose

A good SOP should include a clear purpose. This should be written using language that explicitly identifies the intended aim. Use precise language that your reader could underline to find the purpose.

This section often also includes the scope that is covered by the SOP, whether it is the processes or the individuals for whom the SOP is applicable.

Example:

> This SOP outlines the required security protocols for maintaining the integrity, confidentiality, and availability of Acme organization's IT infrastructure. It aims to protect against unauthorized access, data breaches, and other potential security threats.
>
> This policy applies to all employees, visitors, and contractors who access ACME's IT infrastructure.

## 4 Procedures

The standard operating procedures form the main content for your SOP and should be subdivided according to specific areas that are covered. Make sure that key areas are content-labelled, and that you use logical formatting and alignment to provide a clear hierarchy. Nested listing is common.

Specific tasks should follow the guidelines of procedural writing. Use command voice and more formal, objective language.

## 5 Quality assurance/quality control

Another important part of an SOP includes the methods used to ensure that processes in your organization are proceeding in the best way possible. This includes clear outlining of the individuals responsible for maintaining and controlling quality, as well as a clear explanation of the review process and frequency. Monitoring quality involves ensuring that your SOP is up-to-date, so regular reviews should be scheduled.

## 6 References

SOPs include various reference materials, include specific mention of industry-related standards and guidelines, legal and regulatory documents, organizational policy documents, and technical reference materials and/or academic research publications.

In the case of workplace procedures, this may include context-specific criteria, for example, how frequently security features like passwords must be updated or the use of multiple-factor authentication systems.

Some other elements that may be relevant and required in an SOP include

- Definitions of terms: technical and legal terminology should have clear definitions.
- Description of roles and responsibilities

## 6.6  Review

The following questions will help you consolidate your knowledge about end-user documentation and deepen your understanding through practice.

### Reflection questions

1  What kinds of product or service come with a quick-start guide?

2   When was the last time you used a self-access tutorial to complete a task? What was your experience? Did you have any frustrations?
3   Where are important policies for your organization stored? How are they formatted?
4   What are some of the security policies of the organization you are associated with? What is the password policy?
5   How is your organization's access control organized?

### Application tasks

1   **Identify the end user for each of the following personal mobile device applications in** Table VII.

*Table VII* Types of system admin documentation

| Application | Intended End User |
| --- | --- |
| Health tracking device | |
| Mobile banking application | |
| Online collaboration software tools | |
| Cloud-based storage application | |
| Social media application | |
| Learning management software platform | |
| Language learning application | |
| Health services application | |

1a   What would each user likely already know about the area?
1b   What would each user likely not already know about the area?
2   **Following are the instructions for making the pomodoro recipe listed on page two.** Apply the principles of effective instruction writing to make the recipe more appropriate for an end user.

-----------------------------------------------------------------------------------------------

If using fresh tomatoes, blanch them. You can achieve this by bringing a large pot of water to a boil. Mark an 'X' on the bottom of each tomato by breaking the skin with a knife and drop them into the boiling water for about half a minute. Remove the tomatoes by using a slotted spoon and immediately plunge them into a bowl of ice water. This will make it easier to peel the tomatoes. Once the tomatoes have cooled, peel the skin off and roughly chop them. If using canned tomatoes, skip this step.

In a large skillet or saucepan, heat the oil over medium heat. When the oil is glistening, add the minced garlic cloves and the red pepper flakes. Let the garlic sauté lightly for up to two minutes or until the garlic is fragrant and is just beginning to change colour to golden. Do not let the garlic burn!

If using fresh tomatoes, add them to the skillet along with any juices that have accumulated. If using canned tomatoes, crush them by hand as you add them to the skillet. Combine well using a wooden spoon and then add the dried herbs. Season with salt and pepper according to your preference and stir to combine everything evenly. Turn the heat to low and let the sauce gently simmer on the stove for around half an hour, making sure to stir occasionally. Keep the sauce uncovered, as this will allow water to evaporate and the sauce to thicken, resulting in a greater mix of flavours. If using fresh basil, wait to add it until the last few minutes of cooking.

Taste the sauce to adjust the seasoning as necessary. If you prefer a smoother consistency, you can blend the sauce using an immersion blender. Once it is ready, remove from heat. If using Parmesan cheese, stir it into the sauce until melted.

Pomodoro sauce is delicious in many classic Italian dishes, including pasta, pizza, and chicken Parmesan. It can also be added to roasted vegetables to create a rich base for ratatouille.

---

**2b** Use a voice recognition application to dictate how to make a family recipe of your own. Make improvements to the transcript so that it follows good instructions formatting.

**3  Go to your organization's FAQ page.**

**3a** How is it organized? Which principles does it follow?

**3b** Are there any improvements you could make using the guidelines on page 16?

**4  Design a network topology map for your office or workspace.**

**4a** What devices are connected? How?

## Works cited

1 G. Torkzadeh and W. J. Doll, "The place and value of documentation in end-user computing," *Information & Management*, vol. 24, no. 3, pp. 147–158, 1993. https://doi.org/10.1016/0378-7206(93)90063-Y

2 F. Ganier, "Factors affecting the processing of procedural instructions: Implications for document design," *IEEE Transactions on Professional Communication*, vol. 47, no. 1, pp. 15–26, 2004. https://doi.org/10.1109/TPC.2004.824289

3 P. V. Anderson, *Technical communication: A reader-centered approach*, 9th ed. Cengage Learning, 2018.

4 A. S. Pringle and S. S. O'Keefe, *Technical writing 101: A real-world guide to planning and writing technical content*, 3rd ed. Scriptorium Publishing Services, 2009.

5 A. K. Massey, J. Eisenstein, A. I. Anton, and P. P. Swire, "Automated text mining for requirements analysis of policy documents," in *21st IEEE International Requirements Engineering Conference (RE)*, Rio de Janeiro, Brazil, 2013. https://doi.org/10.1109/RE.2013.6636700

6 J. Bhatti, Z. S. Corleissen, J. Lambourne, D. Nunez, and H. Waterhouse, *Docs for developers: An engineer's field guide to technical writing*. Apress Media, 2021.

7 A. Wallwork, *User guides, manuals, and technical writing: A guide to professional English*. Springer, 2014.

8 D. Both, "Document everything," in *The linux philosophy for sysadmins and everyone who wants to be one*. Apress, 2018, pp. 381–394. https://doi.org/10.1007/978-1-4842-3730-4_20

# Chapter 7

# Report writing in IT

## 7.1   Introduction to report writing in IT

Whatever your job in IT, you will almost certainly need to write reports of different kinds. Many engineers consider them incidental to their real job, so what is the point of them?

### The value of reports

Reports fulfil a number of different functions. They are used to propose, evaluate, recommend, and/or describe the progress of a specific aspect of your job. If we step back from this, though, we can see that their main function is to let people know what you are doing, what you want to do, or what you think should be done.

They are, then, a formal means of informing people about your work. These people are very often members of the management team who lack your technical expertise. They do not necessarily understand how well you have written a piece of code or even what you do most of the day. However, they do understand reports, so these are the channels through which your work interfaces with the decision-making parts of your organization, and where you can help yourself be appreciated.

### Different kinds of reports

Even when we appreciate the value of reports, we might not be sure how to begin writing one. After all, there are many different types [1]. Which one is appropriate? Table I lists some of the most common types of reports and their definitions.

As you can see here, different kinds of reports are used in the lifeline of any project, beginning with the proposal of an idea, the recommendation of an idea, the feasibility of it, through the progress of its development, and finally to an evaluation of its implementation. What this means is that there is overlap between the different kinds of reports. Indeed, an important thing to understand in this regard is that the types of reports are not always distinct. This means that the naming conventions for reports are not always very strict. What one person may mean by a feasibility report, another may think of as a proposal, and so on.

It is therefore more useful to think of the overall purpose of a report. When you are asked to write a report, you are being asked to present information from a specific point of view. You could be presenting much of the same information, but you are framing it as a recommendation, or as proposal, etc. These frames can be reduced to single questions as follows:

- *Proposal*: how can an issue be solved?
- *Recommendation*: which is the best option?

DOI: 10.4324/9781032647524-7

*Table I* Types of reports

| Report type | Definition | Example |
| --- | --- | --- |
| **Proposal** | A proposal is a persuasive document that outlines the technical requirements and details of a new project or service. You write these to convince the reader to implement the proposed plan or approve the proposed project. | A proposal for developing a cross-platform app for small business users to offer automated appointment bookings for customers. |
| **Recommendation** | A recommendation report examines solutions or courses of action in response to a specific situation. It recommends one of them as the best choice. It is usually written to help management make decisions about the most suitable option. | Recommending the best combination for a full-stack development project, choosing between three different options for the front-end and back-end technologies, for example React and Node.js, Angular and Spring Boot, and Vue and Django. |
| **Feasibility** | A feasibility report is a future-oriented document designed to evaluate the likelihood of success for a proposed project, based on specific criteria. As such, it is a tool that helps decision-makers make informed choices about whether or not to proceed with a project, giving them evidence and reasons with which to make their decision. | A report of the viability and potential profitability of a web app using microservices architecture that communicate through REST APIs. |
| **Progress** | A progress report updates readers on the status of a project. It informs them of how well it is keeping to the budget, deadlines, and expected outcomes. It will advise of any challenges and any changes to what has been promised. | A data migration project update, detailing the number of records migrated, the percentage of completion, the quality and accuracy of the data, and recorded validation. |
| **Evaluation** | An evaluation report is a document that assesses the value or effectiveness of something by examining its performance, outcomes, or impacts. Like a feasibility report, it uses a set of criteria and determines how well it meets them. Unlike a feasibility report, it tends to look backwards at completed projects or things in use. However, it may still provide suggestions for improvement or continuation. | An evaluation of a new security software in preventing unauthorized server access, data breaches, and modifications. |

- *Feasibility*: is a project financially and/or technically possible?
- *Progress*: when will a project deliver what it promised?
- *Evaluation*: how well did a project deliver what it promised?

Bearing these questions in mind, we will now briefly look at each of the report types, before turning to some of the features that successful reports of all types have in common.

## 7.2    Proposals

As we have seen, a proposal is a document which addresses an issue and suggests a way to solve it.

### Audience

A proposal is a persuasive document: it seeks permission from someone else to turn a suggestion into reality. As this description indicates, there is a power dynamic behind most proposals: the writer is trying to persuade the reader that their idea is a good one [2]. The reader is the one who has the power to agree with it or not.

Who is going to have the power to agree to your idea? Most people with power in organizations are in management. This means proposals need to be formal in tone and likely need to be written for non-specialists.

If the proposal is to go outside your department, it is often the case that you will need to get the consent of your immediate superior first, as you will be representing the department.

If you are being asked to write a report, it is probably because you have the technical expertise to be able to understand the situation and recommend a technical solution. It is likely that the main reader of your proposal will ask for technical guidance from another expert reader. You therefore need to strike a balance, ensuring that the proposal is technically sound, but also understandable to people without your background.

A further complication in this equation stems from whether the proposal is for an internal or external audience. An internal audience will likely understand the context from which you write much more clearly than an external client. An external proposal is therefore often much longer by necessity, with more attention paid to the background and details to make your case more compelling.

### Purpose

The purpose of a proposal is to persuade the reader that what is being proposed should be implemented. Everything in the document should serve this purpose.

The information you need to include in a proposal to achieve this also depends on whether it is solicited or unsolicited. Solicited proposals are ones people have asked for, whereas unsolicited proposals are ones they have not asked for. Solicited proposals are usually responses to RFPs (Requests for Proposals). An RFP will typically define the issue they want addressed and how they want it done. The purpose of the proposal is to fully answer the brief, some of which can be extremely extensive, time-consuming, and costly to compile.

Unsolicited proposals are ones which you submit without being asked. You will identify the issue the proposal addresses. In this case, the purpose expands to include persuading the reader that there is a problem which needs solving in the first place.

## Layout

Like the other types of report featured here, proposals have no fixed organization. If a proposal is written in response to an RFP, then the structure of it will usually be decided by the authors of the RFP. If your proposal is unsolicited or is in response to a non-prescriptive RFP, then it may be useful to consider the following organizational components:

- **Executive summary**: *what's in the report?* The executive summary outlines the issue, solution, and benefits of the proposal.
- **Problem**: *what is the problem this is a solution to?* Often found in an introduction, the problem is a description of the specific issue which the proposal addresses. Even when a proposal is small-scale, solicited, and internal, it is always best to define very explicitly what the problem is. This way your reader understands the way you see the issue. If this is not clear right at the very beginning, no matter how good the proposal is, it will not solve the same problem that your reader thinks it is meant to be doing.
- **Solution**: *what is the solution to the problem?* This is obviously a key part of the report as it describes the answer to the problem. It may well also describe the benefits of it, whether they are in terms of profit, time, or operational development.
- **Success criteria**: *how do we know when the solution has been successful?* There needs to be an agreed-upon set of criteria which tell all the project's stakeholders that the project has been a success or not.
- **Project strategy**: *how is the project going to be implemented?* This is where you need to set out the detail of your plan. Points you might feature include

  - The project management approach, for example, agile or waterfall
  - What needs to be done, breaking it down into tasks
  - What resources you will need, including people
  - The roles that will need to be filled
  - The potential challenges you might face

- **Budget**: *how much will it cost?* You should include an itemized budget with benchmarked pricings.
- **Timeline**: *how long will it take?* This is where the report outlines a schedule for the project with key milestones and expected completion date.
- **Team profile**: *who will complete the project?* If this is an external proposal, you will need to detail the credentials of the project team and the people leading it. Do they have the qualifications and experience to do the job?
- **Conclusion**: *what is the story?* In your conclusion you can recap the key points and frame them within the story you are telling of solving a problem.

## 7.3   Recommendation reports

It has been suggested that all reports are essentially recommendation reports [3], so it is clearly an important kind of report to understand. Recommendation reports belong to a category of so-called analytical reports which involve using expertise to interpret data and reach a conclusion [4]. The more specific kind of recommendation report discussed here is used when there are a number of options to choose from when faced with a particular issue.

### Audience

If you are writing a recommendation report, it is likely someone wants the benefit of your specialist knowledge. This means they are a non-specialist. You might therefore find it helpful to translate what is of technical value in your estimation, such as security protocols, into less technical measures of value, such as scores or ratings, when you assess the different options.

### Purpose

The purpose of a recommendation report can be divided into two questions:

1  *What criteria can be used to judge the best option?* This involves establishing standards and benchmarks against which an option can be objectively measured.
2  *Using these criteria, which is the best option?* To answer this, all the options need to be compared across all the criteria.

### Layout

Recommendation reports can be based on in-company templates but often there is no set structure to follow. Some of the most common elements include the following components:

*   **Executive summary**: *what's in the report?* The executive summary outlines the context, purpose, and main recommendations of the report.
*   **Issue**: *what is something being recommended for?* Often found in an introduction, this section describes the context for the recommendations. It details the issue to be addressed and the scope of the report.
*   **Criteria**: *how do we know what the best option is?* There needs to be an objective set of criteria which can be used to identify the most suitable way to meet the need identified in the *Issue* section. The criteria should include all the key constraints on choices, such as budget and time, as well as any assumptions which are factored into the final recommendation.
*   **Options**: *what are the possible options?* This is where you review the different options, describing them in relevant detail.
*   **Evaluation**: *how well does each option meet the criteria?* This is where you need to assess the options from the previous sections in terms of the criteria. This is often done using a comparison chart, and in some cases a weighted comparison chart where different criteria are translated into a score.
*   **Conclusion**: *what are the key points?* In the conclusion you can highlight the winners for each criterion and outline the overall strengths and weaknesses of all the options.
*   **Recommendations**: *which is the best option?* This is where you provide your recommendation and the rationale for it. While the Conclusion is often written in the past tense, the Recommendations should be written in the future tense or at least be oriented to the future [5].

## 7.4   Feasibility reports

A feasibility report examines whether an idea is practical to put into action or not.

## Audience

Typically, a feasibility report will have a lot of readers because it touches upon many areas of an organization, from financial to legal to technical, as well as the employees and publics it will affect. Although they may not have the final say on whether a project is given a green light or not, most of these stakeholders will likely have some input. It is therefore a good idea to determine the possible readership in advance and ensure that the contents of the report accurately answer to their needs, particularly those of the decision-makers [6].

## Purpose

The purpose of a feasibility report is to determine if a project is likely to be a beneficial use of resources [7]. To do this, the report answers two questions:

1 *Is it possible?* This involves looking at several areas, including financial, technical, and legal. Does the project require resources, both monetary and human, that the organization has? Does the technology exist? Are there legal restrictions, such as privacy or environmental laws, which prevent the project from being realized?
2 *Is it worth it?* This question looks at the return on investment of a project (ROI). For example, a company may be able to answer 'Yes' to the first question, but the profit, efficiency, or productivity gains the project is supposed to produce are less, or not much more, than the time, talent, and money which need to be invested in it. In this case the ROI is too poor to make the project feasible.

Depending on the project, feasibility reports will place a greater emphasis on one or the other question. However, both questions need to be answered for it to fulfil its purpose.

## Layout

Feasibility reports can cover a lot of different areas across an organization, some of which, such as legal affairs and risk management, have quite different approaches to assessing feasibility. This means there is not one standard way of writing a feasibility report. Nevertheless, some of the most common elements include the following components:

- **Executive summary**: *what's in the report?* The executive summary outlines the proposed idea which is being assessed for its feasibility, the areas with which it is primarily concerned, and the main judgement of the report.
- **Proposal description**: *what proposal is being assessed for its feasibility?* Often found in an introduction, this section describes the proposal being assessed and the need for it.
- **Obstacles**: *what obstacles need to be overcome to make the proposal happen?* This assesses roadblocks to the existence of the proposal, such as limitations of existing technology, budgetary constraints, and legal restrictions. These speak to the first question underlying feasibility reports – is it possible?
- **Benefits and drawbacks**: *how would the proposal benefit and cost the organization?* If the proposal can be done, the question then becomes *should* it be done. Here the pros and cons are weighed up across a lot of variables which all basically assess the opportunity cost: is this the best way for the company to use its resources?

- **Recommendation**: *is the proposal feasible?* This is where you provide your recommendation and the rationale for it.
- **End matter**: *what evidence did you use to arrive at your conclusions?* Feasibility studies are often lengthy documents because they cover a lot of ground and have to demonstrate the evidence, such as benchmarks and test results, which they used to arrive at their recommendation. More so than other kinds of report, you are likely to need to include appendices and reference lists which enable your readers to follow the trail of evidence themselves.

## 7.5   Progress reports

Amongst the other kinds featured here, progress reports display the biggest range in levels of formality as they can often be quite brief.

### Audience

This is because progress reports are often written for your immediate superiors. This is to keep them up-to-date with whatever work you are doing. You may therefore be asked to write small status updates every week or two. More formal, lengthier progress reports are more often appropriate for larger projects which are overseen at a more senior level. Shorter, but still formal progress reports are more appropriate for clients looking to stay informed about projects you are working on for them. The key is to focus on answering the questions your reader is most interested in knowing the answers to [6]. These could be about the timeline, the budget, the effectiveness of the project, and so on.

### Purpose

The purpose of a progress report is to share with stakeholders how well you are doing on a piece of work. This includes outlining the milestones or important developments that have been reached, where the project is in relation to expected progress, and any challenges or obstacles to that progress the project has encountered or expects to encounter [7].

### Layout

Progress reports vary widely in terms of the length and tone they adopt. Nevertheless, some of the common features they include are the following:

- **Project description**: *what are the details of the project?* Often found in an introduction, the description is a useful reminder for the audience and for you about the aim of the project, and what it involves. It can also include details of the client, team members, and agreed completion date.
- **Project status**: *what progress has been made? Is the progress what was planned?* This forms the bulk of the report, and it details what has been achieved so far, whether all or certain parts are on schedule or not, and whether the timeline remains the same.
- **Project challenges**: *what issues are confronting the project?* If there are challenges which may affect the completion or the scope of the project, then they should be identified clearly to avoid surprising your clients and managers.
- **Conclusion**: *what is the overall view of the project?* This summarizes the general outlook for the project, letting your readers know whether everything is going as promised or not.

## 7.6    Evaluation reports

An evaluation report is very similar to a feasibility report but seen from the other end of a project. It looks at whether an idea was practical to implement or not.

### Audience

Typically, evaluation reports are written for management looking to decide the effectiveness and profitability of particular projects, and even of larger strategies. Clients may also ask for evaluation reports, particularly for outsourced projects.

### Purpose

The purpose of an evaluation report is to determine if a project was a beneficial use of resources. To do this, the report answers two questions:

1 *Did it work?* Did the project achieve what it was supposed to? This involves comparing the expected outcomes with the actual ones. If it did not meet or exceeded them, what were the factors that caused this?
2 *Was it worth it?* This question looks at the ROI of the project. For example, if a company was able to answer 'Yes' to the first question, were the profit, efficiency, or productivity gains the project produced worth the time, talent, and money which needed to be invested in it given the context of things as they are now. If so, the project was worth it.

### Layout

Some of the most common elements for evaluation reports include the following components:

- **Executive summary**: *what's in the report?* The executive summary briefly outlines the project that is being evaluated, the context of its use, and the main judgement of the report.
- **Project description**: *what project is being evaluated?* This section describes the project being evaluated, including fundamental aspects like the costs and other resources involved. If it is particularly technical, this is where the key technical aspects can be explained in non-specialist terms. This section can also outline the expected deliverables, outcomes, and targeted metrics (quantifiable measures) of the project. It may present all this in the form of a narrative timeline, documenting the progress of the project to date.
- **Assessment**: *what targets did it meet and miss?* This compares what the project was supposed to do with what it actually did, explaining deviations along the way. This section also considers the change in context from when the project was proposed to its realization sometime later. What were the unknown factors which changed the progress and outcomes of the project? Did it have unforeseen consequences, good or bad?
- **Evaluation**: *was the project worth doing?* This is where you provide your evaluation and the rationale for it. If the project is ongoing, should it be kept as it is, modified in some way, or ended?
- **End matter**: *what evidence did you use to arrive at your conclusions?* As with feasibility studies, you are likely to need to include appendices and reference lists which enable your readers to follow the trail of evidence themselves [8]. You are judging the value of people's work, and therefore there is a lot at stake, so comprehensive evidence is important.

## 7.7    Common report features

As you can see from the previous descriptions of the organizational components of different report types, they have a lot in common. In this section, we look at some of these common elements in more detail.

### Introductions

Why do reports need introductions? Essentially, it's because we want the reader to approach the report with the same understanding that we have. This understanding usually includes several key areas coming from the following five questions:

- *What is the purpose of the report?* Your reader needs to know why they are reading the report and what the report is trying to do. If your audience thinks they are reading a proposal, they will have certain expectations about the kind of content they think should be in it. However, if you have written a progress report instead, but just not made that clear at the start, you will alienate them. It is therefore helpful for everyone if you begin your report with its purpose. You can very explicitly say *The purpose of this report is to propose/recommend/analyze the feasibility of . . .*, etc. That way everyone understands what you are trying to do.
- *What is the context of the report?* In answering this question, you are explaining the background to the contents of the report. Without a proper understanding of the context, the value and meaning of a report are lost. For example, a proposal for a mobile phone will look innovative in 1983, but not in 2023. To help the reader grasp the significance of what you are saying, it is therefore helpful to consider the following questions:
  - What led up to this point?
  - Where is this happening?
  - Who are the stakeholders?
  - When is it happening?
  - Why is it happening?
- *What are the key definitions in the report?* This does not necessarily mean defining key words, although it can involve that too, but rather in its broadest sense of explicitly stating what you want the reader to understand. For example, if one of the key concepts your reader needs to understand is *stack overflow*, ensure that you distinguish between the different meanings and ensure you do so in a way that a non-specialist will be able to follow.
- *What are the parameters of the report?* Another question your introduction should answer is to define its limits. Again, this is as helpful to you as it is to the reader. You need to decide what you will *not* look at as much as what you will. If you are writing a feasibility report, for example, you may want to state that your appraisal is limited to the proposal's technical feasibility rather than anything else. By doing this, you manage your reader's expectations, as well as keep yourself focused on a specific area.
- *What is in the report?* In the final part of your introduction, it is helpful to answer this question, particularly if it is a long report. This overview should briefly tell your reader what they can expect in each section, helping them identify and giving them easy access to the parts that they need or want to read.

Perhaps surprisingly, you may want to write the introduction last. This way you know what is coming in the rest of the report, and therefore you know what areas need defining and explaining for your reader to understand what you have written. Clarity is the key to a successful report, and it stems from thinking about things from your reader's point of view. The introduction is where you can establish that by guiding the reader to the same level of understanding you have.

## Criteria and parameters

The need for clarity and explicitness is not just important in the introduction, but throughout your report. This is particularly true when it comes to criteria.

### What are criteria?

Criteria are standards by which things can be evaluated, judged, and decided. Whenever you make a decision, knowingly or not, you use criteria. If you are deciding what shirt to buy, for example, your criteria might include price, colour, material, style, and local availability.

### Hold on, criteria are . . . ?

Yes, *criteria* is the plural form of the word, where there are more than one. The singular form, when there is only one, is *criterion*.

### OK. What do I need to remember when choosing criteria?

When establishing criteria, it is useful to consider the following points:

- *Criteria must be explicit.* This allows your reader to understand why you have made the judgement you have when you accept a proposal or evaluate the success of a project. Understanding *why* makes your report more persuasive.
- *Criteria do not need to be exhaustive.* No list of criteria covers everything. This is because there is usually no need to be obvious. For example, all laptops have screens, but what differentiates one screen from the other are its size, pixel density, nits, and so on.
- *Criteria must be meaningful and relevant to whatever it is you are evaluating. Relevance* and meaning are often determined by the community of expertise to which you belong, for example, software engineers. This community will have established criteria for judging tools, concepts, and common projects. This makes it easier for you to decide what criteria to use. However, this must be balanced with the next point.
- *Criteria must be focused on your audience's needs.* Context is everything, and your audience's needs are that context. For example, while it is usually lowdown on a list of laptop criteria, brightness is a relevant criterion if a screen is going to be used in daylight a lot. Typically, laptops are primarily used inside under artificial lighting, making screen brightness a far less important consideration. However, outside a dull screen can leave the laptop almost unusable and so it becomes vital if that is where it will be used the most.

### Once I have my criteria, do I need anything else?

Yes, you need *parameters*.

### What are parameters?

To return to choosing a shirt, once you have the criteria, the question then becomes what price, colour, material, and style? And what constitutes local availability? These questions are answered by parameters. Parameters are the values or range of values with which you can measure the criteria. In the case of price, for example, you might have in mind a maximum amount beyond which you are not willing to pay. That amount is the parameter for price.

### But some parameters don't have numbers, so how are they measurable?

That's correct. For example, colours are not numbers, so specifying *blue* as a parameter for choosing a shirt can be a binary value – *yes* or *no* – or it can be graded in terms of shades. Other qualities might need a more subjective grade. Style, for instance, is not something that can be objectively measured, so you might choose a sliding scale: very stylish, quite stylish, not stylish at all. In each case, you can assign a numerical value to help you reach an objective decision. You can compile these in a weighted decision matrix.

### What's a weighted decision matrix?

A weighted decision matrix is a table in which you can compile criteria and parameters to help visualize the decision-making process. Criteria are weighted according to their value. In Table II we can see a simplified weighted decision matrix.

*Table II* Weighted decision matrix for choosing a shirt

| Criteria | Weight | Parameter | Shirt A | Shirt B | Shirt C |
|----------|--------|-----------|---------|---------|---------|
| **Price** | 0.5 | ≤$10 | 1 | 4 | 2 |
| **Colour** | 0.3 | Dark blue | 5 | 2 | 3 |
| **Material** | 0.1 | Cotton mix | 5 | 4 | 4 |
| **Style** | 0.1 | Quite stylish | 3 | 1 | 4 |
| | | **TOTAL** | **2.8** | **3.1** | **2.7** |

This is how the matrix works:

- **Criteria** – the criteria column lists the standards with which we are going to judge the shirt – price, colour, material, and style.
- **Weight** – the weight column indicates how important each criterion is relevant to the others. In this case, price is by far the most important consideration, followed by colour, with material and style equal to each other in last.
- **Parameter** – the parameter column are the values being used to measure the criteria. In this case the range of values for price is $10 or less. This value is then expressed as a number between 1 and 5 (5 = a $2 shirt, 4 = a $4 shirt, 3 = a $6 shirt, and so on). Even non-countable qualities such as colour have been converted to a numerical scale. So here, 5 = perfect dark blue, 4 = lapis, 3 = azure, and so on.
- **Total** – the score for each parameter is multiplied by the weight of the criteria. For example, in terms of price, Shirt A scores 1*0.5=0.5, Shirt B scores 4*0.5=2, and Shirt C scores 2*0.5=1. When all criteria are scored and weighted, the results look like this:
  - *Shirt A* = 0.5+1.5+0.5+0.3=2.8
  - *Shirt B* =2+0.6+0.4+0.1=3.1
  - *Shirt C* =1+0.9+0.4+0.4=2.7

In this case, then, the best suggestion would be Shirt B, although all of them seem close in terms of overall scores. Two points to note about the total:

- The highest possible score is 5 (2.5 + 1.5 + 0.5 + 0.5 = 5) but you or your company may decide there is a threshold below which no option is worth recommending. In this case, for example, you might say you are not prepared to spend money on any shirt below 3.5.
- The way you present the numbers can have an effect on perception. While all the shirts here seem to score similarly when they are measured out of 5, when they are represented as percentages, the relative differences become more apparent:

  - *Shirt A* = 56%
  - *Shirt B* = 62%
  - *Shirt C* = 54%

The value of a weighted decision matrix is that it not only makes decision-making more balanced, crucially it makes it easier to understand for your reader, and more clearly shows them how the decision has been arrived at.

## Money management

Money is like time in that there is rarely enough of it, so it has to be accounted for carefully. However, many of the types of report we have looked at here look forward. That means they require you to guess what money you will need. You can do this by using an *analogous estimation*, which is where you look at the cost of similar projects, or a *parametric estimation*, where you look at the cost of individual resources and add them up. The estimation becomes a *budget* when a *reserve* is added, for example 5% of the total. This is for contingencies.

Whatever method you choose, for smaller projects it is common practice to use tables with itemized costs. This does not need to be complex, but it does need to show exactly what the money will be spent on. Table III shows an example.

While the table itself is relatively straightforward, the figures for each component need to be referenced, with quotations and benchmarking. The numbers in the expense column refer to the relevant endnotes. There also need to be explanations of each item for non-specialists.

## Rhetorical support

As we saw in Chapter 1, we can make writing more persuasive by using the four principles of rhetoric: *ethos*, *logos*, *pathos*, and *kairos*. In Table IV we can see what these mean in terms of writing proposals.

*Table III* Estimate for integration testing project

| Component | Expense | Unit cost | Number | Total |
|---|---|---|---|---|
| Testers | Salary[1] | 1500 | 2 | 3000 |
| Tools | Licence(s)[2] | 1000 | 1 | 1000 |
| Environment | Setup[3] | 500 | 1 | 500 |
| Test cases | Design[4] | 200 | 5 | 1000 |
| Execution | Run[5] | 1000 | 1 | 1000 |
| Reporting | Review[6] | 500 | 1 | 500 |
| Defects | Fix[7] | 500 | 2 | 2000 |
| | | | **TOTAL** | 9000 |

*Table IV* Rhetorical appeal in reports

| Rhetorical appeal | Meaning | How to establish this in reports |
|---|---|---|
| **Ethos** | Ethos is your sense of credibility. It is what makes the reader believe you and have faith in what you write. This is particularly important when you are writing about technical matters to a non-specialist audience. | • Write professionally. Present a report which looks like a report. It suggests that you know what you are doing.<br>• Use and explain key technical terms and processes. This will show you understand your subject and that you care to communicate effectively with your audience.<br>• Proofread for mistakes. Having an error-free report shows you are careful and deliberate.<br>• Insert an *About Us* section, detailing your experience and qualifications in the area. |
| **Logos** | Logos is your use of reasoning and evidence. Without it, your argument will feel ungrounded. | • Provide evidence of key claims. Facts, figures, testimony, mini case studies, and successful, benchmarked examples will all help make your case more persuasive. If the reader can see other similar instances of the same thing, then they are more likely to want to enjoy the same success.<br>• Use logic. As we have noted elsewhere in this book, the notion of causality is very important when persuading people of your point of view. *We should do this because . . . If you do this, then this will happen. This is the situation; therefore, we need to do this.* If the reader can understand why something should be done, they are more likely to do it. |
| **Pathos** | Although pathos is usually used to refer to the emotional appeal of your work, it can also signify the way in which it appeals to the values and interests of your audience. | • People like technology when it makes their job easier, more efficient, and provides a less stressful user experience. Remember that whatever you create, there will likely be a human at the end of it and that they will prefer to feel good about being there.<br>• Security, privacy, cost effectiveness, and productivity are key values which most firms subscribe to, but be sure you know which are important to your client, and mention those. |
| **Kairos** | Kairos refers to your sense of timing and your situational awareness. | • Make sure your report is relevant. Refer to the current situation of your client, and factor in broader technological and macroeconomic trends.<br>• Ensure you are up-to-date with everything in the report, from the pricing to the technology.<br>• Understand the history of your client, even if it's another department in your company. What came before? What is the vision moving forward? |

## Executive summaries

What is the scarcest resource most people have at work? Time. This can be an issue when it comes to reports. Most reports are multi-page documents with different sections and technical details. They are written for readers who are often overwhelmed with reading material which they need to get through in order to make appropriate decisions. What is the solution to this issue? In many cases, it is the *executive summary*.

An executive summary is usually a one to two page summary and overview of a report which comes at the beginning of a lengthier document. As that description suggests, it has a dual purpose:

- *Summary* – it should outline the key points of the document, including the purpose, the context, the key areas, and conclusions reached. In many cases, the intended audience of a report is only one part of the final audience because other stakeholders may wish to understand the contents too. In this case, the executive summary is extremely helpful in providing secondary stakeholders with a sufficient grasp of the key issues.
- *Overview* – in doing the aforementioned, the executive summary also acts as a preview of what the report contains, helping the reader decide which are the key parts they need or want to read.

As you can see, while it may be the last part of a report you write (which is why it is the last part of this section), it is often the first part that gets read, and sometimes the only part. So, what should be in it?

The contents of an executive summary vary according to the report they are a summary of. However, there are some conventions worth considering:

- **Purpose** – define the reason for the report. This involves being clear about the following two questions:
  - *What kind of report is it?* Is it a proposal, recommendation, feasibility, progress, evaluation, or other type? You do not have to define the genre of the report, but you do have to tell the reader what it does. The purpose should, of course, answer to the reader's needs.
  - *What issue provoked the report?* What is the project in the report a solution to? This question essentially asks you to provide the reader with a context for them to understand the meaning and value of the report.
- **Key points** – this is where you need to put the most important issues discussed in your report. For example, if it's a proposal, this is where you describe it. Highlight key data, use diagrams, and illustrate timelines and budgets.
- **Conclusions** – what are the key takeaways for your readers? This is where you put recommendations and final evaluations.

You are not limited to three sections, but you do need to keep the summary brief for it to be of value. Make the information as accessible as possible by organizing it using subheadings, lists, call-out boxes, and other means. Finally, there is a temptation to write an overview in general terms but be as specific as possible. You are not writing about details, but you are writing in a detailed way.

At the top of this section, we mentioned that executive summaries are a very useful way to solve the issue of readers with limited time. Another way is simply to reorder the components of your report, using an executive organization. This is where you front-load the report, putting the conclusions, evaluations, and recommendations at the front of the report instead of at the end. This way busy readers can skip straight to the most important sections and read the rest if they need to. Such an approach is not as efficient as an executive summary, but it does offer a partial solution to the same problem.

## 7.8   Language focus in report writing

So far in this chapter we have been looking at the bigger organizational structures which make up reports. In this section we zoom in and attend to some smaller issues.

### *Maintaining the information flow*

It is more persuasive for your reader if the text of your report runs smoothly. One way you can do that is to maintain the information flow. What does that mean?

The information flow refers to how a text moves from one topic to another. In English, we generally put old information at the beginning of a sentence, and new information at the end. The new information in that sentence then becomes the old information in the next, and so on. Take a look at this simple example:

> It has a terabyte of disk space. This is the minimum size for this job. It requires a lot of data storage.

In the first sentence, *It* is old information because it refers to something we must already know. The new information is that *it* has a terabyte of disk space. As such, it comes at the end of the sentence.

This becomes the old information in the next sentence when it is called *This*. Note that the old information starts the second sentence. The new information is that a terabyte is the minimum size for this job. That comes at the end of the second sentence. This new information – *it* – becomes the old information which begins the next sentence.

This *old – new/old – new/old – new* structure is the natural flow of information in English. Now look at the same sentences written without information flow.

> It has a terabyte of disk space. The minimum size for this job is a terabyte of disk space. A lot of data storage is required for this job.

Here, *a terabyte of disk space* is repeated in the second half of the following sentence as if it were new information, but it is not. Similarly, *this job* is put in the new information part of the third sentence. These misplacements are jarring in English as the expectation that there will be new information at the end of each sentence is thwarted.

It is not always easy to manipulate the information flow, but one way to do it is to use the passive voice. The passive voice is the counterpart to the active voice. Most communication is written in the active voice, for example *He ate the sandwich*. However, we can convey the same information in the passive voice, for example *The sandwich was eaten by him*.

There are several advantages to using the passive voice, including the fact that we can use it when we don't know, don't care, or don't want to reveal who did something. In this case, for example, we can simply say *The sandwich was eaten*.

Another advantage is that it changes the order of the information. Instead of having the sandwich in the new information slot at the end of the sentence, in the passive it is in the old information slot at the beginning. This is a feature we can use to maintain information flow.

We can see an example of this in the following two pairs of sentences. In pair 1, both sentences are active, and both of them end on the phrase QA engineers. However, in pair 2 the second sentence is written in the passive and so the new information from the first sentence can become the old information of the second sentence:

1  *Active – Active:* the company employs QA engineers. When it tests new programmes, the company uses the QA engineers.
2  *Active – Passive:* the company employs QA engineers. They are used when the company tests new programmes.

In this way, the information flow is maintained, and the reader is carried along smoothly by the prose structure.

This old/new information flow can also work at a paragraph level. If, for example, you are providing a list of reasons to support a proposal, and each reason gets a short paragraph, then the beginning of each paragraph should return to the old information (why the proposal is good) before moving to the new information (the next reason why it is good):

- The first reason why this proposal should be supported is . . . X
- Another reason why this is a beneficial proposal for the company is . . . Y
- A third reason to support the proposal is . . . Z

Keeping the information flow at both paragraph and sentence level makes your report easier to read and therefore more persuasive to the reader.

## Conveying stance

The reports we have looked at in this chapter all ask that the writer conveys an opinion to the reader. This opinion is called stance. While stance can be conveyed by explicitly saying something is good or bad, it can also be reinforced more subtly, and help anticipate that explicit statement. How can this be done?

Stance can be conveyed by using the following textual elements:

- *Hedges* – these are words and phrases which show how cautious you are about what you say. Words like *possible, could,* and *generally* indicate to the reader that you are not 100% committed to what you say, that its truthfulness may not be proven to your satisfaction, or that it is not correct in all cases. The most common categories of hedges include the following:

  - Some modal verbs (e.g., can, could, may, might, should)
  - Some modal adverbs (e.g., all in all, arguably, possibly, probably, perhaps)
  - Vague language (e.g., sort of, kind of, somewhat, somehow)
  - Approximations (e.g., about, around, nearly, almost)

- *Boosters* – these are words and phrases, such as *obviously, clearly,* and *of course* which indicate the certainty of a statement. The most common categories of boosters include the following:

  - Some modal verbs (e.g., must, will, shall)
  - Some modal adverbs (e.g., definitely, certainly, undoubtedly)

- Intensifiers (e.g., very, extremely, absolutely)
- Emphatic phrases (e.g., in fact, of course)

- *Attitude markers* – while hedges and boosters indicate the writer's intellectual attitude towards the information in the report, attitude markers show the writer's feelings. There are a number of ways this can be done, including the following:

  - Attitude verbs (e.g., agree, like, prefer)
  - Adverbs which modify whole phrases (hopefully, unfortunately)
  - Evaluative adjectives (astounding, spectacular)

Using hedges can make you seem modest, demonstrating your awareness that you do not know everything. They also work to engage the reader, asking them for their judgement to decide for themselves if something is 100% correct or not. The value of this approach can depend on the cultural context of the audience (see Chapter 3).

Boosters, in contrast, indicate a shared context. You only use a phrase like *of course* when you are confident that the reader will feel the same way. If the reader does not, they will be alienated by what you say, and the report will lose some of its persuasiveness. Too many boosters can make you seem arrogant as a writer, so use them carefully.

Attitude markers should also be used with caution so as not to appear too emotive. This will undermine your ethos as a credible analyst. However, they, like boosters, should not be ignored altogether. Using them to reinforce the main message of your report is a very effective approach, particularly if your report mainly uses hedges. This works like a lexical Von Restorff effect (see Chapter 2), directing the reader to your main takeaways.

## 7.9    Review

The following questions will help you refresh your knowledge about report writing and deepen your understanding through practice.

### Reflection questions

1  What is the value of report writing for both writer and reader?
2  What are the defining features of the intended readers of most reports?
3  Why would you be writing a proposal?
4  What two questions define the purpose of a recommendation report?
5  What two questions define the purpose of a feasibility report?
6  Why might the length and tone of progress reports differ more than other reports?
7  What two questions define the purpose of an evaluation report?
8  Why would you write about the parameters of a report in the introduction?
9  What do weighted decision matrices help with?
10  How can you establish your authority when writing a report?

### Application tasks

1a  The following are the sections of a feasibility report. You need to compile two versions. Place the parts in the right order for a standard and an executive edition in Table V.

- Appendices
- Benefits

*Table V* Sections of feasibility reports

| Standard report | Executive report |
|---|---|
| 1 | 1 |
| 2 | 2 |
| 3 | 3 |
| 4 | 4 |
| 5 | 5 |
| 6 | 6 |
| 7 | 7 |
| 8 | 8 |

- Challenges
- Conclusions
- Costs
- Introduction
- Recommendations
- Summary

**1b** You need to write the executive summary for this report. Which sections would you summarize for it? What order would you put them in?

- Appendices
- Benefits
- Challenges
- Conclusions
- Costs
- Introduction
- Recommendations

**2a** You are writing a recommendation report for the management team. They have decided to replace the office desktop monitors to make screen-based work more comfortable. They have given you no criteria. Which of the criteria in Table VI do you choose, and which do you reject?

*Table VI* Criteria choices for recommendation report

| Criteria | Use? | | Why? |
|---|---|---|---|
| Monitor colour | Accept? | Reject? | |
| Panel type (IPS, VA, TN, etc.) | Accept? | Reject? | |
| Flicker-free technology | Accept? | Reject? | |
| Dullness | Accept? | Reject? | |
| On/off buttons | Accept? | Reject? | |
| Memory size | Accept? | Reject? | |
| Viewing position | Accept? | Reject? | |
| Suitable for visuals | Accept? | Reject? | |
| Connects to CPU | Accept? | Reject? | |
| Good for outdoor use | Accept? | Reject? | |

**2b** Choose a reason why you have accepted or rejected a criterion from the following list by adding the appropriate number in the *Why?* column in Table VI:

1 *Criteria must be explicit.*
2 *Criteria do not need to be exhaustive.*
3 *Criteria must be meaningful and relevant to whatever it is you are evaluating.*
4 *Criteria must be focused on your audience's needs.*

**3a** Change these *Active-Active* constructions to *Active-Passive* to keep the information flow. The first one is done for you as an example.

1  *Active – Active:* one of the selling points of the company were the development and operations teams. The company integrated the development and operations teams.
   *Active – Passive:* one of the selling points of the company were the development and operations teams. They were integrated.
2  *Active – Active:* to meet the business requirements they have to define the data flows. They list the triggers, frequency, direction, volume, and transformations of the data flows.
   *Active – Passive:* to meet the business requirements they have to define the data flows.
3  *Active – Active:* they used an analogous approach to calculate the estimate. They benchmarked against seven similar projects to calculate the estimate.
   *Active – Passive:* they used an analogous approach to calculate the estimate.

*Table VII* Stance verbs

| Stance | Reporting verb |
|---|---|
| Fully agrees with source | 1 Evidences |
| | 2 |
| | 3 |
| Mostly agrees with source | 1 |
| | 2 |
| | 3 |
| Neutral about source | 1 |
| | 2 |
| | 3 |
| Mostly disagrees with source | 1 |
| | 2 |
| | 3 |
| Fully disagrees with source | 1 Claims |
| | 2 |
| | 3 |

**4a** In order to convey your stance about different sources you are using in your report, you need to use the appropriate reporting verbs. Place the following verbs in the right part of Table VII according to what they signify.

- Agrees
- Argues
- Asserts
- Assumes
- Claims

- Conjectures
- Contends
- Declares
- Evidences
- Maintains
- Outlines
- Proves
- Reports
- States
- Suggests

**4b** Choose a verb to complete these sentences about a proposal you wish to recommend.

- The proposal _____ that these efficiency gains mean its budget will be 10% lower than industry benchmarks.
- This software, the proposal _____, will enable integration across all of our company's data input programmes.

**4c** Choose a verb to complete the same sentences about a proposal you do *not* wish to recommend.

- The proposal _____ that these efficiency gains mean its budget will be 10% lower than industry benchmarks.
- This software, the proposal _____, will enable integration across all of our company's data input programmes.

## Works cited

1  G. J. Alred, C. T. Brusaw, and W. E. Oliu, *Handbook of technical writing*. Bedford St. Martins, 2012.
2  M. Chan, *English for business communication*. Routledge, 2020.
3  L. Yeung, "In search of commonalities: Some linguistic and rhetorical features of business reports as a genre," *English for Specific Purposes*, vol. 26, pp. 156–179.
4  J. Balzotti, *Technical communication: A design-centric approach*. Routledge, 2022.
5  S. Mort, *Professional report writing*. Routledge, 2017.
6  P. V. Anderson, *Technical communication: A reader-centered approach*, 9th ed. Cengage Learning, 2018.
7  R. Elling, B. Andeweg, J. de Jong, C. Swankhuisen, and K. van der Linden, *Report writing for readers with little time*. Noordhoff Uitgevers, 2012.
8  P. Hartley and C. G. Bruckmann, *Business communication*. Routledge, 2007.

# Chapter 8

# IEEE referencing and formatting

## 8.1   Introduction to IEEE

This section looks at what IEEE is, who uses it, and why.

### What is IEEE?

*IEEE* (pronounced I-triple-E) is the acronym which stands for the Institute of Electrical and Electronics Engineers [1]. It is an international society for the technology and engineering professions. As well as maintaining a portfolio of standards, distributing and promoting technical information, and creating student and professional networks, IEEE keeps a style guide which is used by most specialists in technical fields, such as IT and computer science. The style guide (which is often referred to simply as IEEE) lays out a standard for the presentation of technical documents, describing how to set out textual elements, like headings, tables, diagrams, and, perhaps most famously of all, references.

### Who uses IEEE?

You can find IEEE in the workplace, academia, and research laboratories. It is most often used by students, researchers, and professionals working in the fields of computer science, IT, telecommunications, electronics, and engineering generally. If a piece of work, such as a proposal or white paper, requires proper referencing, then IEEE is the most popular format to achieve this in these fields.

### What is the value of using IEEE?

As we can see, one of the key benefits of using IEEE is that you are demonstrating that you belong to and understand the presentation standards of the professionals in your field. This immediately gives your work a sense of ethos (see Chapters 1 and 7), letting your readers know that *you* know what you are doing.

Using IEEE to reference the sources for your ideas also provides credibility to what you write in a number of other ways:

- It shows your reader that you have completed an extensive research process. While this is no guarantee of quality work, it does indicate that you are not making everything up.
- It illustrates how other researchers, colleagues, and organizations support the points you are making in your work. This helps validate what you say.

DOI: 10.4324/9781032647524-8

- It demonstrates that you are ethical. When you give credit to others, you show you are not willing to pass off other people's work as your own. This is a testament to your honesty. Further, it helps avoid plagiarism charges at college, and intellectual property theft claims at work.
- It highlights your original ideas. If you reference everything you have been inspired by, what is left is more clearly yours and you should receive the benefit for it.

In sum, then, all the aforementioned *and* the care and attention required to follow IEEE properly helps establish you as a thorough professional. This is always a good reputation to have.

## 8.2    The basics of referencing

This section explains what a referencing style is, what it does, and what kinds there are.

### What is a referencing style?

A referencing style is a way of writing citations and reference lists. It does several things:

1 It indicates that an idea, fact, or point did not come from the author.
2 It tells the reader where the idea, fact, or point *did* come from.
3 It performs both of these functions in a uniform manner across the text so that the reader always knows both 1 and 2.

There are different referencing and document formatting styles. Like coding languages, each of them has their own syntax or structure, all doing 3, but some combining 1 and 2, and some doing them separately. However, at a fundamental level, they all operate in similar ways. There is always a reference in the body of the text which points the reader to the full information in a separate section, either at the bottom of the page or at the end of the text.

Typically, the reference in the body of the text is called a *citation*. The full bibliographical information at the end of the text is called a *reference* and is located in a *reference list*. In this way, each referencing style is like a hash function, using the key of the citation to produce an index as output in the form of the reference list.

### How many referencing styles are there?

There are many referencing styles, and a few very popular ones. These are in use both at university and in the workplace. They include the following types:

- *APA, Harvard*, and *MLA* – these are author-date systems and are variations around citations that look like this – (Shvedhera, 2020).
- *Chicago, Turabian,* and some legal referencing styles, such as *Bluebook* and *OSCOLA* – these use footnotes and/or endnotes in the body of the text to point to the bibliographical information at the foot of the page.
- *IEEE* and *Vancouver* – these use a numerical citation system which point to a numbered reference list.

It is not always the case that technical organizations use IEEE, just as not all medical organizations use Vancouver. Another popular version of IEEE, for example, is IEEEtran (which runs from a set of macros for LaTeX). However, it is the most popular standard in IT and computing generally. Always check with the organization for whom you are writing.

## 8.3   In-text citations

This section looks at citations in IEEE, including how to create them, where to put them, what the different types are, why you would use the different types, when to cite, and when not to.

### How do I cite?

IEEE has the most straightforward citation system of any of the major referencing styles [2]. Citations are shown by numbers in square brackets that look like this [1–3], etc. Unlike other systems, you do not mention names or dates.

There are several important points to note about this:

- Every [#] refers to a separate source. You cannot use one number to reference a group of sources.
- The numbers refer to the order in which the citations come in the text. [1] is the first, [2] is the second citation, and so on.
- If you refer to the same source more than once, use the number you originally gave to it, even if that is out of sequence. Do not create duplicate citations.
- If you need to refer to multiple sources in the same place, use the following protocol:

  - For an unbroken series of citations in terms of numbers, use a dash to separate the first and last references, for example [6–9]. This refers to [6–8] and [9].
  - For a broken series of citations, use commas to separate the references, but keep them in their own brackets, for example [4–7, 11].
  - It is also possible to use a combination of the two types of series, for example [3], [5–7, 12].

### Where do I cite?

Every citation should be situated on the same line as the text, before any punctuation (such as periods/full stops/commas/colons), with a space before the bracket. Look at the following examples:

- . . . following the example of IBM in the 1980s [3].
- . . . following the example of IBM in the 1980s [3], a model which later inspired many.

When you place a citation at the end of a clause, it indicates that what went *before* it is attributable to the author, but what comes *after* is not. If the citation comes at the end of the sentence, it is all attributable to the author. Look at the following examples:

1 The full-stack skillset will soon be necessary for all web coders [17], but likely not for another decade.
2 The full-stack skillset will soon be necessary for all web coders, but likely not for another decade [17].

In example 1, the citation covers only the first phrase, but not the caveat at the end about not for another decade. In contrast, the citation in example 2 comes at the end of the sentence and therefore covers both phrases.

### What are the different styles of citation?

As with all referencing styles, there are two styles of citation – *parenthetical* and *narrative*.

**Parenthetical citations** are probably the type of citation we think of first when we think of citations. Sometimes called *non-integral* citations, they are, as this name suggests, not part of the grammar of the text. They look like this:

- In comparison, Groffen was using an inefficient array in its data structure [43].
- Agile lead times call for an incremental approach [5], a less rigid task-based approach [19], and constant testing [32–35].

As we can see in these examples, the citation makes no difference to the meaning and grammar of the sentence. If we remove it, the sentence essentially remains the same.

**Narrative citations**, on the other hand, cannot be removed. As their other name, *integral* citations, indicates, this type of citation is part of the grammar and the meaning of the sentence. Take a look at these examples:

- In comparison, as [43] contends, Groffen was using an inefficient array in its data structure.
- According to [38], hash tables are too inefficient for this.
- [5] states that agile lead times call for an incremental approach, while [19] adds that they also need a less rigid task-based approach.

Here the citations work as proper nouns. They substitute for the authors' names. For example, [5] = Yadav – *Yadav states that agile lead times call for an incremental approach.* Removing the citations would make the sentences almost meaningless. *Who* exactly states that agile lead times call for an incremental approach?

When you substitute a citation number for a noun, such as a name or a pronoun, it is good practice to use the name the first time you use the citation. Have a look at this example:

- Yadav [5] notes that a dev-ops approach allows for more disruptive innovation. In contrast, [5] states that agile lead times call for an incremental approach.

In the first sentence, Yadav is mentioned as the author of [5]. Having done this, the next sentence is able just to use [5] as a substitute for the name.

### Why do I use the different styles?

The simple answer to this depends on the kind of *stance* you want to display in your text. As we saw in Chapter 7, stance refers to the way in which you convey your attitude about what you are saying. In the case of citations, you are letting the reader know what you think about your source or what the source is stating.

The difference between parenthetical and narrative citations is that parenthetical citations do not require you to demonstrate a stance, while narrative ones allow you to do so more easily if you wish. Compare and contrast the parenthetical and different narrative versions of this statement:

1  Educational technology is the next most important revenue generator [25].
2  [25] *states* that educational technology is the next most important revenue generator.
3  [25] *proposes* that educational technology is the next most important revenue generator.

4  [25] *claims* that educational technology is the next most important revenue generator.
5  [25] *conjectures* that educational technology is the next most important revenue generator.

Sentence 1 is the normative version of the statement, simply recording what the text says. However, once the citation is put in a noun position (before the *verb*), there is no option but to indicate your stance towards the statement, even if that stance is a neutral one as it is in sentence 2. However, as we progress through sentences 3–5, the sceptical attitude becomes more pronounced until it is clearly hostile in sentence 5.

As was mentioned in Chapter 5, it is more effective if strong stance markers like the one in sentence 5) are used more rarely. This means that your default option should probably be to use parenthetical citations, reserving narrative ones for when you wish to make an impact.

### When do I cite?

The simple answer to this is that you cite your source every time you either quote from them or paraphrase their work.

A quotation is when you copy from the source text directly. A paraphrase is when you put it in your own words. Compare and contrast the following two examples:

Direct quotation
Development is a gradual process that 'can act only by very short and slow steps' [2, p. 510].
Paraphrase
The development process consists of a series of small changes over a long period of time [2].

As you can see, a direct quotation is indicated by using single quotation marks. It also requires a page number from the source. A paraphrase requires neither.

If your quotation is approximately three lines, about 40 words, or longer, then you are advised to indent it from both margins as a block quotation, and reduce the font size, like this:

> There are several reasons for the failure of big tech to anticipate the effects of AI on the marketplace. The first was that much of the development was undertaken behind closed doors. Development teams from different companies not only did not speak to each other but were legally not allowed to [15, p. 62].

While it is obvious when to cite a quotation, the same cannot be said for paraphrases. This is because almost everything everyone says is an idea or thought that someone else has had at some stage before.

### When do I not cite?

You usually do not use citations when what you are saying is *common knowledge* or you are referring to *your own opinion* (although there is an important exception to that), or when you are still using the same source already cited.

#### Common knowledge

Common knowledge refers to information that most people know, such as Python is a coding language. It also refers to information your reader is likely to know. This depends on their context. If they work in the same company or field as you, you will have a larger pool of shared knowledge which does not need to be referenced.

Another feature of common knowledge is whether your reader can argue with what you are saying. Over the last decade, the field of accepted facts has shrunk to the point where the Earth is considered flat by some. You will likely know your readers and what they presume to be givens, so you will know what needs to be supported by sources. A final consideration is if what you say is supported by sources. If your reader can easily find half a dozen good sources to support what you say, then it is common knowledge.

### Citing yourself

Generally, you do not need to cite yourself. However, the big exception to that is in academia, particularly when you are a student. If you have submitted something elsewhere which you then recycle for another assignment, you need to cite yourself if it is short or paraphrase it. This is because the exact same wording will be flagged by software universities use to detect plagiarism.

### Continuing a citation

If you cite a source and continue to discuss the idea from it, do you have to use the citation number in every sentence? No, you do not but only if you continue to indicate by other means that the ideas come from the same source.

As we mentioned at the beginning of 7.2, the first two functions of citations are to let the reader know that an idea is not yours, and to tell them whose it is. Once you have inserted a citation, you can continue to fulfil these two functions by other means. You can use phrases that refer the same source. Take a look at this example:

Galloway [6] proposes that technology is valued for its ability to save time more than anything else. He advises us that medical care is the next industry ripe for disruption. Currently, vast waiting times are built into the client experience. He argues that AI will likely change that.

All four sentences here refer to the same source, but only the first one has a citation. The second and fourth sentences use a pronoun – *He* – to refer to the citation in the first sentence. This way we as readers understand that all the sentences, including the sandwiched third sentence are paraphrasing Galloway.

However, if that third sentence refers to another source, we have to reintroduce the Galloway citation in sentence four:

Galloway [6] proposes that technology is valued for its ability to save time more than anything else. He advises us that medical care is the next industry ripe for disruption. Currently, according to Lin [7], vast waiting times are built into the client experience. [6] argues that AI will likely change that.

This underlines the importance of understanding the principles behind citations, and not just memorizing the rules.

## 8.4 End-reference list

Every in-text citation you use needs to correspond to a full reference in a list at the end of your main text.

### How do I create a reference list?

There are several points to note when making your reference list:

- The list should be ordered by number, not by alphabet. It should begin with [1], the lowest number, and progress in sequence from there.
- The number of citations you have should correspond to the number of entries in your references list.
- Every citation must be in the reference list.
- No reference should be repeated. If you refer to a source multiple times, it still has only one entry in the reference list.

This is an example of the first three entries of a reference list:

[1]  S. Goericke, *The future of software quality assurance*. 2020. doi: 10.1007/978-3-030-29509-7.
[2]  One World Initiative Broadcaster, "Scott Galloway: Disrupting higher education," *YouTube*. Jan. 02, 2022. [Online]. Available: www.youtube.com/watch?v=1kBfZOtWqgU
[3]  A. Kaur and K. Kaur, "A COSMIC function points based test effort estimation model for mobile applications," *Journal of King Saud University – Computer and Information Sciences*, vol. 34, no. 3, pp. 946–963, Mar. 2022, doi: 10.1016/j.jksuci.2019.03.001.

As you can see here, IEEE reference lists are formatted in a clear and accessible way. It involves using the following features:

- Put the citation numbers in their square brackets next to the left-hand margin.
- Use a tabbed space from the number to the reference entry.
- Entries should be single spaced.
- There should be a double space between entries.

The next step is to understand the syntax or layout of each kind of entry.

### What does each entry need in a reference list?

As with most referencing styles, the reference list contains all the information that anyone needs to find the same source themselves. In order to be able to do this, each entry should include

- Author
- Title
- Date
- Source

Let's look at these in turn before turning to specific types of reference.

## Authors

Unlike other referencing styles, IEEE puts an author's first name first, but only the first letter, like this:

*Table I* Author names in IEEE reference lists

| Number of authors | Example |
| --- | --- |
| 1 | T. Pincawan, |
| 2 | T. Pincawan and P. B. Uwir, |
| 3–6 | T. Pincawan, P. B. Uwir, and P. Sinurat, |
| 7+ | T. Pincawan *et al.*, |

As we can see in Table I, names are fairly straightforward in IEEE. Please note the following points:

- There is a period/full stop after each initial.
- There is a comma after each author's surname *except* the first name when there are two authors, *and* after the first name when there are seven or more names.
- If there are seven or more authors only write the first author's name and then write *et al.* afterwards. This is a Latin abbreviation meaning *and others*. It needs a period/full stop after *al.* because this is abbreviated from *alia*. Put a comma after that as you would normally.
- Authors can be organizations or corporate authors as they are sometimes called. Give the name of a corporate author in full.
- If there is no author, use the title of your reference instead. Write it in full.

## Titles

The next piece of information you need to include is the title of your reference. In IEEE, like other referencing systems, the formatting depends on whether the title is of something part or whole:

- "Titles of parts" – a source is part of something if it is contained by something bigger. For example, a chapter is part of a book, and an article is part of a journal. The logic is not always clear (for example, company reports are considered parts), but generally it is.

  - Titles of parts are enclosed in double quotations marks, like this – "Levels of efficiency gain in dynamic arrays".
  - Titles of parts are written with sentence capitalization. This means that the first word of the title is capitalized, as is the first word after a colon, and of course names, but nothing else, like this – "Levels of efficiency gain in dynamic arrays: The pointer arithmetic coefficient".

- *Titles of wholes* – a source is whole if it contains smaller units inside it. For example, a book is a whole because it has chapters inside it. Likewise, a journal is a whole because it has articles inside it.

  - Titles of wholes are written in italics, like this – *Handbook of Software Quality Assurance*.
  - As you can see from this example, the titles of wholes are written in title capitalization. This means that every word apart from the small words, like articles and prepositions, are written beginning with a capital letter.

## Dates

IEEE referencing includes two types of date: the date of publication and the date of access.

- *Dates of publication* are used for when a source was made. These are generally used for non-mutable texts (texts that do not change), such as published books, journals, reports. For sources published only once, only a year is required. For sources published multiple times in a year, such as a journal, months and years are required.
- *Dates of access* are used for when you took the information from something that may change, such as a website. In order to pinpoint the date of access as closely as possible, days, months, and years are required.

## Sources

Sources refers to the origins of a reference. This covers a lot of different information depending on the reference, but it can include any of the following:

- The whole of which a reference is a part, such as the book for a chapter, the journal for an article, or the website for a page.
- The publishers for books, companies and organizations for reports, occasions for presentations.
- The locations of publishers, companies, organizations, and occasions (for example, conferences), such as cities and states.
- The addresses of URLs and DOIs. DOIs are digital object identifiers which act as permanently assigned addresses for academic work.

As we can see, this is a lot of information to think about for each source. Luckily, you won't need all of these for every reference because different types of sources require different kinds of information. We will look at in these in the next section.

## Referencing internet sources

In IEEE, you can reference websites as in Table II:

Table II  Referencing websites in IEEE

| | |
|---|---|
| Format | Initial(s) of First Name. Last Name. "Webpage Title," Website Title. URL (Accessed Month Day, Year). |
| **Example** | K. McKelvey. "The best software developer blogs to read," *We Are Developers.* www.wearedevelopers.com/magazine/software-development-blogs (accessed Aug. 03, 2023). |

There are several points to note here:

- IEEE has several suggested formats for online references. The aforementioned is the most basic one.
- Unlike with other IEEE references, you need to put a period/full stop after the author's last name, *not* a comma.
- There is no punctuation after the URL.
- The month is almost always abbreviated – see 7.5.

## Referencing reports

In IEEE, you reference technical reports as in Table III:

*Table III* Referencing technical reports in IEEE

| | |
|---|---|
| *Offline format* | *Initial(s) of First Name. Last Name, "Title of report," Abbrev. Name of Co., City of Co., Abbrev. State, Country, Rep. xxx, year.* |
| **Example** | S. Jónsdóttir, "A hash tree protocol," RLM, Reykjavík, Iceland, Tech. Rep. 51556-12-RLM, 27 Oct. 2001. |
| **Online format** | Initial(s) of First Name. Last Name, "Title of report," Abbrev. Name of Co., City of Co., Abbrev. State, Country, Rep. xxx, year. Accessed: Abbreviated Month Day, Year. [Online]. Available: URL |
| **Example** | S. Jónsdóttir, "A hash tree protocol," RLM, Reykjavík, Iceland, Tech. Rep. 51556-12-RLM, 27 Oct. 2001. Accessed: Jan 21, 2024. [Online]. Available: rlm.org/rep/sjons5155612RLM |

## Referencing standards

In IEEE, you reference standards as in Table IV:

*Table IV* Referencing standards in IEEE

| | |
|---|---|
| **Offline format** | *Title of Standard, Standard number, date.* |
| **Example** | *Python Parameter Coefficients*, Python Standards, 2004. |
| **Online format** | *Title of Standard*, Standard number, date. [Online]. Available: www.url.com |
| **Example** | *Python Parameter Coefficients*, Python Standards, 2004. [Online]. Available: www.pythhonstandardsassociation.com |

## Referencing articles

In IEEE, you reference journal articles as in Table V:

*Table V* Referencing articles in IEEE

| | |
|---|---|
| *Offline format* | *Initial(s) of First Name. Last Name, "Article title," Journal Name, vol. #, no. #, pp.#-#, Abbreviated month Year, DOI.* |
| **Example** | M. M. Zhou, "Software is hardware", *Engineering Principles*, vol, 18, no. 2, pp. 35–52, Jan. 2022, doi: 10.1080/10310600.2022.738738. |
| **Online format** | Initial(s) of First Name. Last Name, "Article title," *Journal Name*, vol. #, no. #, pp.#-#, Abbreviated month Year, DOI. [Online]. Available: URL |
| **Example** | M. M. Zhou, "Software is hardware", *Engineering Principles*, vol, 18, no. 2, pp. 35–52, Jan. 2022, doi: 10.1080/10310600.2022.738738. [Online]. Available: www.engprin.org/10310600/2022/738738 |

There are a couple of minor points to note here:

- If the reference ends with a DOI, then it should be completed with a full stop/period.
- If the reference ends with a URL, then it should not be completed with a full stop/period.

## Referencing books

In IEEE, you reference books as in Table VI:

*Table VI* Referencing books in IEEE

| | |
|---|---|
| *Offline format* | *Initial(s) of First Name. Last Name, Title of Book. City of Publisher, (only U.S. State), Country: Publisher, Year* |
| **Example** | P. Wash, *AI, AI Captain: Augmenting Naval Navigation*. Cambridge, MA, USA: NIT Press, 2025 |
| **Online format** | Initial(s) of First Name. Last Name, *Title of Book*. City of Publisher, (only U.S. State), Country: Publisher, year. Accessed: Abbreviated Month Day, Year. [Online]. Available: site/path/file (database or URL) |
| **Example** | P. Wash, *AI, AI Captain: Augmenting Naval Navigation*. Cambridge, MA, USA: NIT Press, 2025. Accessed: Apr. 1, 2026. Available: https://nitpress.com/nav/23998530 |
| **Offline chapter format** | Initial(s) of First Name. Last Name, "Title of chapter in the book," in *Title of Published Book*, Initial(s) of First Name. Last Name of editor, Ed., City of Publisher, (only U.S. State), Country: Publisher, year, ch. #, pp. #–#. |
| **Example** | D. Rafiq, "Almma head out: Automatized meme generation," in *Generative E-Commerce*, P. Didd, Ed., London, UK: Nodoubtledge, 2023, ch. 12. pp. 175–189. |
| **Online chapter format** | Initial(s) of First Name. Last Name, "Title of chapter in the book," in *Title of Published Book*, Initial(s) of First Name. Last Name of editor, Ed., City of Publisher, (only U.S. State), Country: Publisher, year, ch. #, pp. #–#. Accessed: Abbreviated Month Day, Year. [Online]. Available: site/path/file (database or URL) |
| **Example** | D. Rafiq, "Almma head out: Automatized meme generation," in *Generative E-Commerce*, P. Didd, Ed., London, UK: Nodoubtledge, 2023, ch. 12. pp. 175–189. Accessed: Oct. 5, 2024. [Online]. Available: https://nodoubtledge.com/genecom/ch12/44895 |

These are just some of the most common forms of references you will likely need to use. As you now know the main components of a reference and how they are presented, it is likely you will be able to cite most forms of references well. What we choose to reference is changing rapidly, and this is not always reflected in the referencing systems which were built to reflect a world of published work. Different forms of social media, for example, are not always accounted for. What IEEE does is provide you with the principles you can use for whatever forms of reference have yet to be incorporated into their system.

## 8.5   Abbreviations and locators

This section looks at IEEE abbreviations and alternative locators.

## List of common abbreviations

As we have seen already, IEEE referencing uses a lot of abbreviations [2–3]. One of the most common sets are the months of the year. These are mostly reduced to three letters, but some are kept at four:

- Jan. for January
- Feb. for February

- Mar. for March
- Apr. for April
- May for May
- June for June
- July for July
- Aug. for August
- Sept. for September
- Oct. for October
- Nov. for November
- Dec. for December

Of the four-letter abbreviations, June and July do not require periods/full stops after their use because they are complete words, but Sept. still does. Of the three-letter months, May does not need a period/full stop.

Other common abbreviations IEEE permits the use of include the following:

- *Comp.* for computing
- *Conf.* for conference
- *Dept.* for department
- *Inf.* for information
- *Rep.* for report
- *Tech.* for technical
- *Technol.* for technology
- *Univ.* for university

One abbreviation IEEE does not let you use is *ibid.* which stands for *ibidem*, which is Latin meaning *in the same place*. This is a common abbreviation in other referencing styles but don't use it in IEEE.

### List of common locators

Sometimes you want to cite something that does not have page numbers. This is particularly true for electronic and online sources, but it can also happen with other media, such as presentations. When it does, you can use alternative locators – other forms of address for the information you are citing.

We can see some examples of ways to do this for in-text citations in Table VII:

*Table VII* Common locators

| Locator | Example |
| --- | --- |
| **Equations** | . . . as is clear in [4, eq. 1] |
| **Figures** | . . . as displayed in [2, Figure 8] |
| **E-Books** | . . . as suggested in [2, Chapter 3] |
| **Online Article** | . . . as evidenced by [17, para. 4] |
| **Tables** | . . .as seen in [4, Tab. 11]] |
| **Unnumbered Pdfs** | . . .. as contended by Banjo [2, "Nutrition" section] |
| **Videos** | . . . as shown in [1, 03:26] |

As with all referencing systems, there is not an exhaustive list for every eventuality. The question to ask yourself is whether someone unfamiliar with the source could find it from your directions alone. If not, then you need to make them more precise.

## 8.6    IEEE style guide

As mentioned at the beginning of this chapter, IEEE is not just a way of formatting references, but also a complete style guide for your work [4]. If you are submitting to an IEEE-affiliated organization, there are a number of rules you will need to follow, such as writing in two columns. However, in less strict environments, attention is generally paid to a few areas, such as tables and figures.

### Tables

Table VIII is an example of a table formatted in IEEE style (apart from the font choice).

### Title

As you can see, every table begins with a title which announces it as a table. It is numbered sequentially in Roman numerals, for example III, IV, V, VI. This is written in block or upper case capitals. The title of the table is written below this in lower-case capitals or sentence capitalization. Both are centred above the table itself. There should be a space between the title and the table. The title should not be part of the table.

### Lines

The table should be divided by horizontal lines, although these can be omitted, if necessary. Use vertical lines only when it is required to reduce confusion to the reader.

### Alignments

Numbers are usually aligned on the decimal point if there is one. Other than that, alignments within the table are at your discretion.

*Table VIII* Key factors in using tables in IEEE

| Point number | Point |
| --- | --- |
| 1 | Give each table a caption |
| 2 | Centre caption *above* the table |
| 3 | Write TABLE in block capitals |
| 4 | Number using Roman numerals |
| 5 | Write caption in lower-case capitals |
| 6 | No full stop/period |
| 7 | [1]Explanatory notes follow |

Explanatory notes go below the table.

[1] Superscript numbers can be used to refer to specific parts of the table

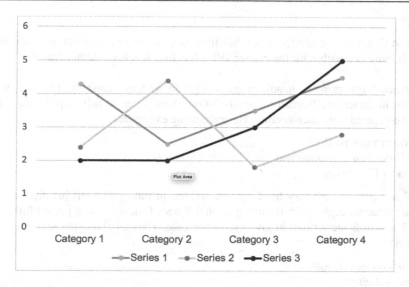

*Figure 8.1* Key factors in using figures in IEEE

*Note.* Explanatory notes go below the figure, with general notes first.

## Figures

Whereas tables are strictly defined as all data presented in the format of a grid, figures have a much wider meaning as they cover everything else, apart from equations, that is not in the text. This means, for example, that if you put graphs or charts in your work, they are all referred to as figures.

The formatting of figures is almost the opposite of tables, as we can see in Figure 8.1.

### Caption

The first way figures are distinct from tables is that rather than titles, they have a caption which goes underneath the figure. The words are written in lower or sentence case. With only the first word capitalized. Captions should be descriptive.

### Figure

The word *Figure* should not be written out in full. Rather it should be written in the abbreviated form as Figure, then leave a space before writing the number in Arabic numerals, for example Figure 8.6.

### Alignment

The caption should be centred under the figure, and ideally appear within the borders of the image. The legend should also be placed within the borders of the figure. Any explanatory notes need to be written below it.

## Headings

As we saw in Chapter 2, headings and subheadings help organize your work, give it consistency, and make it more accessible for the reader. IEEE has four levels of subheadings:

- *Level 1 headings*: primary headings are written in sentence capitals. The section number is written in uppercase Roman numerals followed by a period/full stop. Primary headings should be centred over each section. Here are some examples:

  I. INTRODUCTION TO IEEE
  II. THE BASICS OF REFERENCING
  III. IN-TEXT CITATIONS

- *Level 2 headings:* secondary headings are written in italics and left-justified, not centred. They are ordered alphabetically using capital letters followed by a period/full stop and a space. The words are written in lower sentence case. They sit above the text. Here are some examples:

  *B. List of common locators*
  *F. When do I cite?*
  *D. Income streams*

- *Level 3 headings*: tertiary headings are indented usually by one tab stop from the left margin, (although strictly they should be a one em indentation). They are written in italics and ordered by Arabic numerals followed by a right parenthesis or bracket. The words in the heading are followed by a colon. The text follows the colon on the same level as the heading. Here are some examples:

  *1 One dimensional arrays:* these arrays include . . .
  *2 Two dimensional arrays:* contrary to one dimensional arrays . . .
  *3 Three dimensional arrays:* a third but not final category of arrays . . .

- *Level 4 headings:* quaternary headings are indented twice from the left margin, and also italicized in lower case. They are ordered alphabetically, with letters followed by a right parenthesis. As with level 3 headings, they are separated from the body text by a colon. Here are some examples:

  *a) One dimensional arrays:* these arrays include . . .
  *b) Two dimensional arrays:* contrary to one dimensional arrays . . .
  *c) Three dimensional arrays:* a third but not final category of arrays . . .

There are a couple of points worth noting:

- REFERENCES is a level 1 heading that should not be numbered.
- If you have one appendix, format it like a primary heading but without a number. If you have appendices, they must be numbered but the numbering is not connected to the rest of the paper. You can use Roman or Arabic numerals, and they come after the word, for example APPENDIX 3.

## 8.7   Drafting tips for organizing source material

When you are writing, it is unlikely that your first draft will be perfect. You will probably want to move sections, delete paragraphs, and insert sentences for your second draft. When you do this, it will change the order of your references, meaning that the numbers are no longer in the correct sequence. There are two ways to deal with this: by hand and by software.

### Drafting by hand

Many people use a surname/date referencing system while drafting their papers. That way there is a clear reference to attach an IEEE numbered citation to when the paper is finished, and the final order of the list is clear. It is then easy to use *Find and Replace* on your writing software, replacing the *(surname/date)* with the appropriate *[#]*.

Similarly, you can keep a list of tables, figures, and equations and their titles on a separate document and then apply the correct numbers to them when you have finished, making sure you tick off each one as you go.

Indeed, for both the reference list in the paper, and your list of tables, cross reference them. Is there an in-text citation for every reference in the reference list? Does the number of tables and figures and equations match the ones in your list?

### Drafting with software

There are two types of software available for this – integral and non-integral:

- *Integral referencing software* – these are referencing programmes which are part of your writing app. For example, Word has a whole section devoted to references which can be formatted in IEEE. You can also get add-ons and macros for apps like Google Docs and LaTeX which do a similar job. The advantage of using them is that they change the citation number when the citation is moved around the document. They also change the corresponding number in the reference list. This will save you a lot of time.
- *Non-integral referencing software* – these are referencing programmes which are part of your writing app. Most of these, like Zotero, EndNote, and Mendeley are both hardware-based and cloud-based apps, enabling you to sync and back up across devices, as well as work offline. If you use the plug-ins, they also have the functionality of the writing apps, but in addition they enable you to store references independent of individual documents in a central bibliography. This can be useful if you are likely to reuse references across papers and projects.

Whichever type of software you use, be sure to check for errors. None of them are infallible because they only work with whatever inputs they are given.

### 8.8.  Review

The following questions will help you refresh your knowledge about IEEE referencing and formatting and deepen your understanding through practice.

### Reflection questions

1   What are the key benefits of using IEEE in your work?
2   What three things do all referencing styles do in a text?
3   Why would you use a parenthetical citation style?
4   Why would you use a narrative citation style?

5  When do you cite?
6  When do you not cite?
7  What does every entry in a reference list ideally need to have?
8  Why do we want that information?
9  What is the value of the formatting styles for headings and subheadings in IEEE?
10  Why would you choose to use non-integral referencing software over integral software?

### Application tasks

1  In the following examples, you are provided with both an original text and an example of how it could be paraphrased and used in your text. Choose the best place for the citation in the following extracts – [a], [b], or [c] – to show you are referencing only the idea from the original text.

1a  **Original text**: web coders will eventually need to master the full-stack skillset, which involves both front-end and back-end development.

**Your text**: the full-stack skillset will soon be necessary for all web coders [a], but likely not for another decade[b], and even then, only in small companies[c].

1b  **Original text**: web coders will eventually need to master the full-stack skillset, which involves both front-end and back-end development. However, this might take another ten years to happen, and even then it might only apply to small companies.

**Your text**: the full-stack skillset will soon be necessary for all web coders [a], but likely not for another decade[b], and even then, only in small companies[c].

1c  **Original text**: web coders will eventually need to master the full-stack skillset, which involves both front-end and back-end development. However, this might take another ten years to happen.

**Your text**: the full-stack skillset will soon be necessary for all web coders [a], but likely not for another decade[b], and even then, only in small companies[c].

2  Which of these ideas do you need to support with a reference?
2a  A report which mentions that IT affects all areas of modern life.
2b  A comment about a new form of array that was made to you in a private email.
2c  A new idea in an old TikTok video.
2d  An idea that was quoted by someone else.
2e  An idea in a paper you wrote.
3  Which of these references are parts of a whole, and which are whole?

[1]  P. Narang and P. Mittal, "Performance Assessment of Traditional Software Development Methodologies and DevOps Automation Culture", *Eng. Technol. Appl. Sci. Res.*, vol. 12, no. 6, pp. 9726–9731, Dec. 2022.
[2]  S. M. Stone, "IT equals 'It's tough,'" in *Management for professionals*, 2018. doi: 10.1007/978-3-030-01833-7_7.
[3]  R. Alt, G. Auth, and C. Kögler, *Continuous Innovation with DevOps: IT Management in the Age of Digitalization and Software-defined Business*. Springer, 2021.
[4]  *Artifact Traceability in DevOps: An Industrial Experience Report EASE '23: Proceedings of the 27th International Conference on Evaluation and Assessment in Software Engineering*June 2023Pages 180–183https://doi.org/10.1145/3593434.3593451

**4a**  In IEEE abbreviations, which of these months is four letters and which three?
1 March
2 April
3 July
4 August

**4b**  Which of these are IEEE abbreviations?
1 *Rept.* for report
2 *Techn.* for technical
3 *Technol.* for technology
4 *Uni.* for university

**5**  What is missing from the following references?
**5a**  M. M. Zhou, "Software is hardware", *Engineering Principles*, vol, 18, no. 2, pp. 35–52, Jan. 2022, doi: 10.1080/10310600.2022.738738. Available: www.engprin. org/10310600/2022/738738
**5b**  D. Rafiq, "AImma head out: Automatized meme generation," in P. Didd, Ed., London, UK: Nodoubtledge, 2023, ch. 12. pp. 175–189.
**5c**  K. McKelvey, "The best software developer blogs to read," *We Are Developers*. www. wearedevelopers.com/magazine/software-development-blogs
**6**  Put these heading styles into the correct order, with the primary heading first.

- *F. When do I cite?*
- *3) Three dimensional arrays:* a third but not final category of arrays . . .
- *B. List of common locators*
- III. IN-TEXT CITATIONS

## Works cited

1 IEEE, "IEEE home page." [Online]. Available: www.ieee.org/ [Accessed 26 March 2024].
2 IEEE, "IEEE reference guide," 2023. [Online]. Available: https://journals.ieeeauthorcenter.ieee.org/ wp-content/uploads/sites/7/IEEE_Reference_Guide.pdf [Accessed 26 March 2024].
3 IEEE, "List of IEEE journal/magazine titles, internal acronym, and reference abbreviation." [Online]. Available: https://docs.google.com/spreadsheets/d/1D4xH38098a2xpq0sO16o9gQuje1bq_-GOinRRPTYZac/ edit#gid=0 [Accessed 26 March 2024].
4 IEEE, "IEEE editorial style manual for authors," 2023. [Online]. Available: https://journals.ieeeauthor-center.ieee.org/wp-content/uploads/sites/7/IEEE-Editorial-Style-Manual-for-Authors.pdf [Accessed 26 March 2024].

# Chapter 9

# Multimodal communication in IT

## 9.1 Introduction to multimodal communication

This section looks at what multimodal communication is, who uses it, and why.

### What is multimodal communication?

Multimodal communication is the use of more than one way to communicate at a time. For example, when we go to a movie, we can hear a soundtrack, sound effects, and voices, and we can also see moving pictures. If there were no pictures, for example, it would be unimodal, like radio or podcasts.

As we have seen, typically, IT involves using multimodal resources, such as text and visuals together, each with their own languages which can complement the other.

However, we need to be aware that when we use a different mode of communication, it not only changes how we send it but how we receive it too [1]. In this section we will look at the main ways we communicate, both as senders and receivers of messages, and how those differences affect the messages themselves.

### Difference between speaking and writing

So far in this book, we have mainly looked at written and visual forms of communication. However, a good part of our day is also spent speaking. It is useful to be aware of the differences that shifting between speaking and writing can entail [2], as we can see in Table I:

*Table I* Differences between speaking and writing

|  | Speaking | Writing |
|---|---|---|
| **Time** | Speaking is typically more spontaneous. We make it up as we speak. Signs of this include the use of fillers, such as *um* and *ah*. | Writing is typically more prepared, and we have to edit it before we let others read it. This gives it a more polished appearance. |

(Continued)

DOI: 10.4324/9781032647524-9

*Table I  (Continued)*

|  | Speaking | Writing |
|---|---|---|
| **Non-linguistic cues** | Speaking involves many more non-linguistic cues than written language. We can say the same thing with different intonations, for example, which can make the words mean the exact opposite of each other. Similarly, body language is another way of promoting or undermining a message. | In written language, we are usually restricted to question, quotation, and exclamation marks if we wish to use non-linguistic cues. Increasingly, emojis are becoming popular as a supplementary orthography, although they are generally thought of as informal and so not used in more formal communications. All of which means we need to be more careful when writing than when speaking. |
| **Grammar** | When we speak, we use complete sentences far less than when we write. We also tend to use contractions (for example, didn't, wouldn't, can't etc.) more often. | Written language is often considered more formal. It generally uses full sentences without such errors as run-ons and fragments. |

## Difference between listening and reading

The differences between speaking and writing, as the senders of messages, are also felt by the receivers of messages when we listen and read [3]. Table II outlines some of the differences:

*Table II*  Differences between listening and reading

|  | Listening | Reading |
|---|---|---|
| **Time** | When we listen, we do so instantaneously. We cannot return to what someone said unless it is recorded. This makes listening more demanding than reading. | When we read, we can return to what was said before and re-interpret it in the light of what comes later. Further, the text is permanent so it becomes a memory of what someone communicated which can be consulted at any stage. |
| **Feedback** | As a listener, we can often give the speaker feedback. Even if we cannot say anything ourselves, we can let the speaker know what we think using non-linguistic cues, like folding our arms, smiling, or staring off into space. | Reading is usually asynchronous, so there is no opportunity to let the message sender know what we think when we are reading. This prevents them from altering their message to better suit their audience and can lead to readers feeling resentful and frustrated. |
| **Control** | As a listener it is harder for us to control the flow of information. We have to proceed at the pace of the speaker. Typically, we listen a lot faster than we read. | When we read, we can alter the pace of information to suit ourselves. If something is very technical or uses language we do not know, we can slow down, try to work it out from context, or simply consult a reference guide of some kind. |
| **Memory** | Listening is a greater burden on our memories. Because we cannot go back to connect ideas to what is currently being said, we have to keep it in our minds for the duration of the other person speaking. | The text works as your memory when you read. This means you can use more mental resources on understanding. |

*Multimodality and accessibility*

As was mentioned in Chapter 2, multimodality is a useful way of making content easier to access. This is equally important in the context of technical documentation as work and study spaces become increasingly diverse. Some people prefer interacting through speech, while others prefer using text. Some, like people who don't use English as a first language or people who process texts in a different way, such as those with dyslexia, can benefit from visual cues that support written or spoken content.

*Implications of modal differences*

The implications of these differences mean that how you send and receive information with your colleagues and clients is crucial. For example, when you need to give a presentation about some technical information to a non-specialist audience, not only do you need to simplify your language, but you need to circle back and refer to points you have previously mentioned, as well as possibly give them a handout with highlights and a glossary. You should also pause regularly and ask for questions.

The implications of modal differences also have implications for the medium you choose to communicate in. Returning to your technical presentation, for instance, you might decide that in order to give your audience the benefit of being able to track back and rewatch points, you are going to record it as a video. As another example, you might choose to video conference a client rather than phone them so you can read their non-linguistic communication more easily, and they can read yours. Alternatively, you may opt to write to a client instead so that you can choose what you say very carefully, and not have to make it up on the spot.

## 9.2   Video conferencing

One form of communication which is multimodal is the online meeting. Apps, such as Zoom and Teams, have made video conferencing extremely popular. It is usually cheaper and more efficient than a face-to-face meeting and offers a greater sense of social presence than a phone meeting. Meetings within an organization are more likely to be video conferences, especially if they involve simple tasks and there is little chance of ambiguity. Meetings which require building relationships, such as with external clients are more likely to be face-to-face [4].

However, there are also issues that need to be addressed when video conferencing. Have a look at Table III at possible problems and solutions to them.

## 9.3   Presentations

This section looks at best practice in presentations, examining the audience and purpose of presentations, effective ways to structure your decks, approaches to slide creation which will help your audience understand you, and types of delivery.

### The audience for presentations

Presentations are given to various kinds of audiences. It is essential you tailor your talk to the audience's needs. Failure to do this will result in an ineffective presentation. When you think about audiences, it is useful to consider their likely mood, level of knowledge, and time.

*Table III* Video conferencing issues and solutions

|  | Video conference issues | Solutions |
| --- | --- | --- |
| **Time** | While video conferences are meant to be fully synchronous in the same way face-to-face meetings are, this is unlikely to be the case. Even a half-second lag between participants can lead to a higher incidence of interruptions, people talking over each other, and even unnaturally long silences. | Establish explicit turn-taking ground rules at the beginning of a call. For example, if someone wishes to speak, they can use the hand-raising emoji. If you are having technical issues, let someone know so that it is clear you are not being rude. |
| **Feedback** | Cues are harder to see online. This means that interpersonal connections are harder to form and maintain. | Use more explicit repair strategies than you would face-to-face. To do this, repeat or clarify what someone just said. For example, 'Just so I am clear here, you are saying that. . .' Use emojis to signal agreement and disagreement. |
| **Focus** | Because we are not immersed in our surroundings when we are in a virtual meeting, we can become distracted by our actual surroundings. | To maintain focus, establish a clear agenda at the beginning so that people know where the meeting is going. If people know what the end is and how they get there, they are more likely to stay with you. Check in with all participants regularly. |

These considerations then give us some of the most common types of audience, including the following:

- **Neutral audiences** – they are neither for nor against you.

    - Tip: if people do not know you, focus on developing your ethos through a competent and professional performance which is not overly personal.

- **Hostile audiences** – through no fault of your own, some audiences will not be in a mood to listen to you.

    - Tip: avoid personal anecdotes, jokes, and unnecessary points. Focus on logos as the driving force of your presentation, using plenty of facts and reasons.

- **Non-expert audiences** – you are a specialist in your area, but your job is to demonstrate your expertise in an audience-friendly fashion.

    - Tip: don't try to explain everything. It is better to explain some things well than explain everything poorly.

- **Expert audiences** – they know what you know, at least up to a point, and you can present to them on the same level.

    - Tip: focus on what is new. They already know the context so the value-add for them is the same as it is for you – the original thoughts.

- **Managerial audiences** – they have busy schedules and little time to engage with detail. They have to make decisions, and they therefore need the information they can trust to make them.

  - Tip: keep your presentation brief. Deliver the main takeaways upfront and provide them with an executive summary.

## The purpose of presentations

Everyone has to give presentations of one kind or another. They are an opportunity to offer well-researched thoughts to an audience primed (or not!) to listen. But *why* do we use them?
There are several key purposes for presentations, but they tend to centre on the following:

- **Informing** – giving project updates, providing overviews of services in an area, or outlining challenges and possible solutions for the audience to choose from are all examples of presentations which are designed to inform.
- **Selling** – proposing a service to a client, suggesting a project to your management team, or explaining to your colleagues why one approach is more effective than another are all instances of persuading your audience of a point of view.

In both cases, these presentations require that you engage with your audience and keep them on your side. Indeed, the key advantage presentations have over written reports is that you can get feedback from your audience, answering questions and clarifying understandings [5]. In the next two sections, we will look at how you can keep your audience engaged by structuring your presentation in certain ways, and by creating slides that help your audience to understand what you are saying.

## Structuring presentations

In this section we will look at three of the most common and effective ways to structure your presentation.

### Deliver in threes

The rule of threes has been around for nearly 2500 years, and it still works today. Essentially, it means that audiences will only engage with three main points. While this is not necessarily true, it is seductive. There are a few ways to apply it to your presentation:

- **Structure your presentation around three tentpoles**. This means that you should have three big ideas. If you are presenting a recommendation, for example, you could have a problem, the solutions, and the recommendation.
- **Tell your audience a story with a beginning, a middle, and an end.** As we shall see in the next section, stories are great ways to engage audiences in presentations. In a proposal presentation, you might describe a problem at the beginning, detail the solution in the middle, and finish by outlining the happy ending this will give your potential client.
- **Inform your audience what you are going to tell them, tell them, then tell them what you told them to sum up.** This is a classic structure for a presentation. Your audience will know it and will understand what you are doing. It is always good to manage expectations, and the preview at the beginning lets the audience know where you are going. By summing up at the end and repeating the points, this helps with the challenge of listening and not being able to backtrack.

It is perfectly possible to do all three combined, and many good presentations do. These are guidelines, however, not rules. There are many rules which have sprung up about presentations, such as the 7x7 rule. They are unnecessarily restrictive, and while there are good reasons for their invention, it is what works for the audience which is the most important consideration.

## Tell them a story

Using a story structure is a good way to engage your audience. The most common way of doing this is using the *SCQA* approach developed by Barbara Minto [6], as we can see in Table IV.

At first glance, this may not look like any stories you know, but in fact most fit this formula. Most begin with the normal life of the protagonist (the situation), something happens to change that (the complication), this then poses a question about the protagonist (the question) which they then spend most of the movie or book answering (the answer). People pay billions every year to be entertained, comforted, and intrigued by this structure. It is therefore a very useful tool if you want to engage your audience.

The real key to its success, however, is that it answers the *So what?* question. By structuring the presentation as an answer to a question, you give the audience a reason to listen. Every slide and every diagram suddenly has an angle. It is not a factual presentation, but a presentation of facts in support of something bigger.

## Build a pyramid

Underpinning the SCQA is the Minto Pyramid Principle [6]. This is a very effective way of structuring presentations. The way it works is simply by beginning with your main message, providing arguments to support that, and giving evidence for those supporting points.

The main message is essentially the answer from the SCQA approach. In fact, the two ideas are often used in combination. Let us look at an example, answering the question: *what is your favourite mobile operating system?*

Level 1, the top of the pyramid, is the main idea, answering the question. The reason why that is the answer can be found in Level 2, the second layer of the pyramid, where the supporting ideas are located. The support for the reasons can be found on the bottom of the pyramid in level 3.

The pyramid is the way you organize information in your presentation when planning it. When you give the presentation, you start with Level 1, then a Level 2 point, with the Level 3 supporting points, and then back to the second Level 2 point, and so on.

*Table IV* SCQA in presentations

| Part | Explanation | Example |
| --- | --- | --- |
| **Situation** | This is the context of your presentation. It is where you briefly describe what is happening. | Company is looking to grow revenue with its podcasting app |
| **Complication** | This is the challenge arising from the context. What has happened to change things? | Podcasting apps are many, and differences are few |
| **Question** | This is the challenge framed as a question. How can the challenge be addressed? | How can podcasting app differentiate itself? |
| **Answer** | This is the response to the question. It forms the bulk of your presentation. | Company should develop API to connect with video podcasts |

In each case, the point on the previous level is a summary of the ones connected below it. This means that you have a degree of flexibility when presenting. If you have limited time, you can skip the lower levels. If time is extremely short, you can simply stick to Level 1. This front-loading of presentations, where you give the main idea, recommendation, or conclusion first, is extremely useful if you are presenting to an executive audience with limited time. If they have time, they can watch the whole presentation, but if not, they have still got the gist of what you want to say.

## Slide design

Having looked at the big picture of presentations, let us now focus on the slides you use to deliver them. There are lots of slide templates available, but which should you choose and why? This section begins with a reminder of who your slides are actually for and how the design language of your deck should always be written with them in mind.

### The challenge for audiences at presentations

A lot of people find presentations difficult, but the people who find them the most challenging are your audiences. This is because the challenge of presentations is a multimodal one. At any one time, your audience can be doing several or all of the following:

- **Listening to you.** One of the disadvantages of listening is that you cannot go back and read over what was said, either to try and understand it for a first time, or to revisit a point in the light of later information. Listening to a presentation is a linear activity, and it therefore requires concentration.
- **Reading your slides.** While reading allows you to return and read it again, often the parts you are being asked to read are dense with information and require deciphering, such as graphs and charts.
- **Engaging with your content.** If your presentation is effective, then your audience will think about what you are saying and possibly begin applying it to their own contexts.
- **Wondering how they will respond to you.** If an audience is engaged, they will want to formulate and ask questions, or they may even be questioned by you. If they are surrounded by an audience of peers, this is a testing scenario for many.

This is a lot for our brains to process effectively, so how can the design of your presentation help your audience to do so?

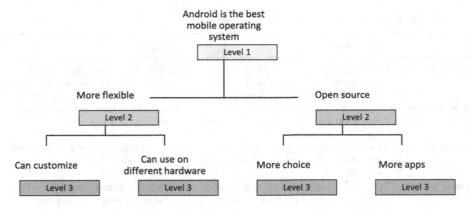

*Figure 9.1* Minto Pyramid Principle example

One solution to this is to keep your slide deck minimal. This is possible in certain scenarios, such as when presenting a product, but is harder in others, such as when your audience needs to be persuaded by research. Equally, a slide deck can live on after a presentation and be used for reference. If you adopt a minimalist approach, this can damage your slides' legacy by making them impossible to understand without you.

What, then, can you do to help your audience meet the presentation challenge? The following sections highlight some of the approaches that can be taken to help your *combined* approach make sense.

## What's the So What?

Facts and data are commonly used in presentations, but they are pointless unless they are used for a reason. Sometimes referred to as the *So what?*, this reason is essentially the argument of your presentation. Facts and data are only there to support your argument. They do not mean anything by themselves.

When you review your slides, ask of each one: *so what?* It may be informative, but does it contribute to the overall argument?

Think of presenting as a team game. A slide by itself may be brilliant, but if it does not contribute to the overall purpose of the presentation, bench it. Having a coherent message tied to a coherent purpose helps your audience meet the presentation challenge much more effectively.

Similarly, if your presentation does not contribute to the overall purpose of the project, then it is pointless.

## Introductions

To get the audience interested in your presentation, you may want to use a *grabber*. A grabber is simply a way of getting people engaged in what you are saying by grabbing their attention. As you can see in Table V, there are different kinds of grabbers, suitable for different kinds of audience.

The purpose of a grabber is not to be interesting in and of itself; rather, a grabber should get the audience interested in the subject of your presentation. For this to happen, it should be clearly connected to what you are going to say. If you use a surprising statistic on global API revenue, for example, then your presentation must be about that too.

## Titles

The title for a slide is usually at the top of it. When we read visually, we start at the top and move our eyes downward. Therefore, anything written at the top of a slide is going to get our attention first. This is your title.

But titles do not stop there. They are also usually written in title-formatted font, that is font which is either the biggest font size on the page or bolded or both. On a slide with text of different sizes, our eyes are naturally drawn to the bigger text first. We will therefore read the title first.

Given both these points, you can see that titles have a natural advantage over any other text on the slide. It should therefore not be wasted, but what is the best use of it? One way to use the titles is to write the biography of the slide. What is the slide's story, the slide's *So what?*

By *biography* we mean the main point of the slide. Just as when someone tells you their life story, they don't tell you everything, just the main idea. With a slide, ask how it contributes to the story, and then use that as the title. Look at these two title options for the same slide:

1  Number of consumers using mobile payment apps over last ten years.
2  395% increase in number of consumers using mobile payment apps over last ten years demonstrates market for micro-payment app.

*Table V* Grabbers in presentations

| Grabber | Example | Audience |
|---------|---------|----------|
| *Stating the content* | 'This presentation will look at the benefits and costs of developing an app for the Android market'. | Expert/Hostile/Managerial |
| *A surprising statistic* | '9999 out of 10000 mobile apps fail to make a profit'. | Expert/Managerial/ Non-expert/Neutral |
| *A short story* | 'The other day, I got a food delivery. It was quick and the food was hot, so I wanted to tip the driver. However, I had no cash in the whole house. What were my options?' | Neutral/Non-expert |
| *A quotation* | 'The best way to get a project done faster is to start sooner'. – Guido Van Rossum | Expert/Neutral |
| *A rhetorical question* | 'Outside of finance, why is healthcare the sector with the biggest investment in blockchain technologies?' | Expert/Managerial |

The difference between these two is clear. In the first one, there is a description of the graphic on the slide. This is accurate, but what's the *So what?* The second title option is also a description of the slide, but there are two key differences:

- **Data are contextualized.** By showing both that there has been an increase and by how much, this title gives much needed context to the chart. Context is what gives things meaning.
- **Chart has purpose.** By connecting the chart to the overall presentation aim, you show the chart's purpose. You give it a reason to be there and a reason for the audience to pay attention.

Titles, then, are not the names of slides; they are the reasons why people should get to know the slides at all.

### Kickers

*Takeaway, call-out* or *kicker* boxes are the text boxes at the bottom of a slide. These boxes are usually formatted in a way to draw the attention of the audience, perhaps with a brighter colour or larger font. As always, there is an opportunity cost here. When you make your audience focus on one area of a slide, they will not be focusing on other areas. So, when is it appropriate to use them?

There are two main reasons for using a kicker:

1 **When what you are saying is not immediately apparent.** For example, if you have a slide which has a complicated graphic on it or a lot of facts and figures, it can help the audience if you direct their attention to the main point of the slide. It could be argued that if your slide is that busy, you should remove it. However, sometimes it is helpful to show your audience dense graphics and charts in order to give them the larger context and to demonstrate the extent of your research. This gives your presentation an ethos boost and makes it more persuasive.

2 **When you want to segue to the next slide.** For example, you might have a question in your takeaway box. The answer to this question can be the topic of your next slide. This is a

semi-visual version of the new-old pattern for sentence structure we looked at in Chapter 3. It makes your slides flow like a story.

Try not to add new information to a kicker. Audiences expect key takeaways in it. If they see new information instead, they think they have missed something and will focus on that and not on what you are saying.

## Charts and chart design

When you make presentations, you will often have to use charts as a way to support your point. Their purpose is to make understanding the relationships between data quicker and easier than can be done with talk or text. To do this, there are two factors to consider: choosing the right type of chart and making an appropriate design. Let's look at each of these in turn.

Firstly, there are many different kinds of charts you can use, but they are all designed for specific purposes. Here are four of the most common:

- **Bar charts**

  - *Purpose*: to compare categories against each other
  - *Tips*: order your bars, for example, from biggest to smallest, always start the X axis at 0 and do not use 3D bars as they distort the data
  - *Useful language*: the most common . . . , the leading . . . , the principal . . .
  - *Example*: the popularity of different wallet apps

- **Pie charts**

  - *Purpose*: to compare categories against a total
  - *Tips*: limit the number of categories to no more than five, ensure that the percentages total 100, put the biggest slice at 12 o'clock and move in order from there
  - *Useful language*: the relative popularity of . . . , the distribution of market share of . . . , the relative prominence of . . .
  - *Example*: the percentage of market share for different wallet apps

- **Line charts**

  - *Purpose*: to show how categories change over time
  - *Tips*: label lines on the right of the chart (rather than use a legend), colour the lines, and highlight key periods
  - *Useful language*: the changing landscape of . . ., the growth/decline in . . . , the evolution in . . .
  - *Example*: the trend in wallet app use over past ten years

- **Scatter plots**

  - *Purpose*: to show the relationship (or lack of it) between variables and to highlight outliers
  - *Tips*: the more data points, the more reliable the data seems, use colour to highlight key data points, add a line to show the trend more clearly
  - *Useful language*: the association between . . . , the correlation between . . . , the link between . . .
  - *Example*: the relationship between wallet app use and cost of phone

Of course, it may be that you do not need a chart at all. While charts are good for visually demonstrating relationships between data, tables are better at highlighting specific values within data ranges. It could be, then, that you should opt for a table instead if that suits your purpose.

When you do use a chart for the right purpose, you need to ensure that the design serves that purpose. Often, this is about avoiding misleading features, such as axes which do not start at 0. However, sometimes it is about avoiding features altogether and not just the misuse of them. For example, 3D pie charts can be extremely misleading because they are made to be looked at from a specific angle which distorts your clear view. In other cases, you will need to remove the clutter, such as excessive labelling (do you need to write the months out in full?), pictures for chart backgrounds, and even gridlines and axes.

Instead, use only texts, colours, lines, and numbers which are absolutely necessary. Further, use the chart title to direct the audience to the value of the chart for your presentation: what is the *So What?*

### Icons

Using professional icons on your slides can help your audience process the type of information more quickly. For example, an icon of a handshake prepares the audience for content about meetings or networking. An icon of a clipboard with ticks on prepares the audience for content about task lists. And an icon of a calculator prepares the audience for content about budgets.

Just to be clear, ClipArt is never acceptable in professional presentations, but icons are.

### Annotations

You can add notes or graphic highlighters to make your diagrams and charts more accessible. Your audience may not be specialists in your area, and they certainly will not know the diagrams you use as well as you.

You can make it easier for your audience in a number of ways:

- By highlighting key parts in different colours
- By inserting notes
- By using highlight boxes set around important areas
- By creating zoom bubbles which exaggerate certain aspects of a graphic

If you adopt all or some of these approaches where appropriate, you will help your audience understand your presentation and be able to engage with it a lot more effectively. Ultimately, then, the challenge for your audience is the challenge for you.

### Distractions

As we have already noted, presentations are very demanding on their audiences. There is no need to add to that challenge with distracting visuals. There are three main culprits here:

- **Backgrounds** – avoid busy backgrounds as they distract from the content of your slide. Indeed, a useful guideline to remember is that if your audience notices your background, then it is already no longer a background. To this end, it is hard to beat a plain-coloured background against which your text and visuals can stand out. A simple white background is often a good choice, although for longer presentations darker backgrounds cause less eye strain for viewers and/or offer a welcome contrast. Try to avoid using any of the templates that come with PowerPoint, Keynote, and other presentation software.

  - **They are too busy.** Most of them draw attention to the design because they were, after all, created by designers who like their work to be noticed. These busy designs also end up using a lot of screen real estate, crushing your content into a smaller space.

- **They are cliched**. Anyone who has even a passing acquaintance with PowerPoint will have seen them before. Using one of them is suggestive of someone who has not made the effort.
- **They have amateur connotations**. Every schoolchild who has ever made a presentation will have used a ready-made template.

- **Visuals** – visuals, such as pictures, dominate attention over text. They therefore need to be used sparingly and only when they have a purpose [7]. This means not only avoiding inappropriate visuals, but any visual that does not make the point better than text could.
- **Colours** – there are a couple of points to consider with regard to colour choice:

  - **Limit the number of them.** As with templates, excessive use of colours can be distracting. It is better to limit yourself to a small harmonious palette.
  - **Use them strategically**. When you do use colours, they draw the eye. This can be used to highlight key points, such as when you colour the important bar on a bar graph.
  - **Create your own palettes.** As with the templates, try to avoid using the pre-programmed colour palettes. They are extremely common, and it looks lazy when you use them. It does not take much effort to create your own palette and for those who lack colour sense, there are many palette creators on the internet.

### Delivery

There are four established styles of presentation delivery:

- **Scripted** – this is where you write out the script of your presentation and read it out.

  - This is good for when you need to deliver a carefully worded message.
  - This is bad for the effect on the listener as eye contact with your audience is limited and delivery can be stilted.

- **Memorized** – this is where you deliver the whole presentation from memory.

  - This is good for when you need to deliver a carefully worded message *and* like to maintain eye contact.
  - Unless you have a photographic memory, this requires a stupendous amount of effort, especially for longer presentations. It also doesn't allow for much audience interaction.

- **Impromptu** – this is where you make an entire presentation up in the moment.

  - This is good for when you have no time to prepare *and* you know the material extremely well.
  - This is likely to result in a failure to deliver the message effectively as you will not have been able to organize the information flow, create a story, or implement the rule of threes.

- **Extemporaneous** – this is a cross between memorized and impromptu. In it, you use notes, which can range from a short outline of the whole talk to bullet lists of keywords and phrases for each slide.

  - Because it is less rigid than a fully scripted talk, this is good for most presentations which require engagement with the audience and the capability to respond to them. Equally, it is more organized than an impromptu presentation, so it should be easier for your audience to follow.
  - It requires time to rehearse and still does not guarantee perfect precision.

The key to all of the delivery methods, apart from the impromptu one, is practice. Only when you run through your presentation several times do you see what works, where the challenges are, such as difficult words to pronounce, and whether you can hit your time limit effectively. Even an impromptu presentation only works if the speaker is practised both at presenting and in their subject matter.

### Checklist for presentations

This is a checklist you can use to ensure you make the best presentation you can:

## 9.4   Language focus in presentations

As we have mentioned already, one of the difficulties for listeners is that they are unable to go back and check where they are in a presentation. Is this still the same point, or has the speaker moved on to another one?

The best way to help the listener understand where you are in the presentation is to use *signposting* language. Like actual signposts which tell drivers which direction they are heading in, these are words and phrases that direct the listener to the part of the presentation you are in. They are not content, but language that contextualizes the content and helps make sense of it.

### SCQR signposts

If you are using the SCQR structure for your presentation, you can let your audience know where you are in the presentation by using some of the following words and phrases:

**Situation**: currently/right now/broadly
**Complication**: however/but
**Question**: therefore/the result of this, this raises the question of . . .
**Answer**: the answer to this is . . ./the most effective solution to this is . . .

*Table VI* Checklist for presentations

| Part of presentation | Question | Yes/No |
| --- | --- | --- |
| **Purpose** | 1 Do you know why you are giving the presentation? | |
| **Audience** | 2 Do you know who you are giving the presentation to? | |
| | 3 Is the presentation designed for that audience? | |
| **Structure** | 4 Have you organized your presentation? | |
| | 5 Does it tell the audience the situation? | |
| | 6 Does it tell the audience the complication? | |
| | 7 Does it tell the audience the question? | |
| | 8 Does it tell the audience the answer? | |
| | 9 Does it front-load the information? | |
| | 10 Does it have supporting points? | |
| **Slides** | 11 Is it clear on each slide what the *So what?* is? | |
| | 12 Does each slide have a title which makes a connection to the purpose? | |
| | 13 Have you used kickers which provide takeaways? | |
| | 14 Have you used charts appropriate to the purpose? | |
| | 15 Have you minimized distractions? | |
| | 16 Have you rehearsed? | |

## Transition signposts

When you move between points in your presentation, it is helpful to the audience if you let them know. This can be done in several ways, including letting the audience know you are concluding one part and starting the next. This doubling up of signposts means that even if a listener misses one signpost, they hopefully won't miss both.

**Summarizing**: all in all, this suggests that . . ./in summary, it is clear that this all means that . . .

**Previewing**: first, I want to talk about . . ./this raises the question of what we should do next . . ./ next, we will move on to look at . . .

**Connecting**: the second point we need to consider is . . ./as promised at the beginning of the presentation, this means we need to turn our attention to . . ./this connects to our previous point about . . .

## Closing language

One area that is often overlooked is giving your presentation an ending. While it is simple to say 'In conclusion . . . , this is not particularly engaging. Rather, it is more engaging to employ some *You* language (see Chapter 3) and to direct the audience towards your key idea.

**Recommending**: what I hope you get from this presentation is the value of . . ./the key takeaway for me is that we need to . . ./I recommend that above all you consider. . .

**Asking**: what I would like you to do next is . . ./given all this, I ask that you choose to . . ./all of which leads me to request that . . .

## 9.5    Review

The following questions will help you consolidate your knowledge about multimodal communication and deepen your understanding through practice.

## Reflection questions

1  Why do you need to be aware of multimodal communication?
2  What are the key differences between speaking and writing?
3  What are the key differences between listening and reading?
4  Why would you video conference rather meet face-to-face?
5  What issues do you need to consider when video conferencing?
6  How do you alter your presentations for different audiences?
7  What does SCQA stand for?
8  What is the advantage of using the Pyramid Principle when structuring your presentation?
9  What's the *So what?*
10  What do you use titles, kickers, icons, and annotations for?

## Application tasks

**1a**  Place the attributes in Table VII in either speaking or writing as appropriate:

- Complete sentences
- Large selection of non-linguistic cues

- Prepared
- Sentence fragments
- Small selection of non-linguistic cues
- Spontaneous

*Table VII* Speaking or writing attributes

| Speaking | Writing |
|---|---|
|  |  |

**1b** Place the attributes in Table VIII in either listening or reading as appropriate:

- Asynchronous
- Higher memory burden
- Interactive
- Less control over information flow
- Lower memory burden
- More control over information flow
- Non-interactive
- Synchronous

*Table VIII* Listening or reading attributes

| Listening | Reading |
|---|---|
|  |  |

**1c**  Given the following priorities, what communication modality would you choose: *spoken* or *written*? Complete Table IX

*Table IX*  Most suitable modality

| Priority | Modality |
|---|---|
| Precise wording | |
| Technical information | |
| High-context cultural cues | |
| Brainstorm | |
| Understanding of non-native speakers | |

**2**  You are hosting a video conference call. What three ground rules would you establish at the beginning of the call?

**3**  What is good advice for which kind of presentation audience? Match the audience type to the advice in Table X.

- Neutral
- Hostile
- Non-expert
- Expert
- Managerial

*Table X*  Audience types and advice

| Audience type | Advice |
|---|---|
| 1 | Avoid personal anecdotes, jokes, and unnecessary points |
| 2 | Focus on what is new |
| 3 | Deliver the main takeaways upfront |
| 4 | Develop your ethos through a competent and professional performance which is not overly personal |
| 5 | Don't try to explain everything |

**4a**  Put the stage of the presentation in the right order to tell a story.

a) Answer
b) Complication
c) Question
d) Situation

1
2
3
4

**4b**  Match the stage of the presentation to the example of the stage by writing the number in Table XI.

*Table XI* Presentation stage examples

| Presentation stage | Example of stage |
| --- | --- |
| | How can podcasting app differentiate itself? <br> Podcasting apps are many, and differences are few. <br> Company should develop API to connect with video podcasts. <br> Company is looking to grow revenue with its podcasting app. |

5   Put the letters corresponding to each point in the appropriate box to illustrate the Pyramid Principle in Figure 9.2.

a) Open source
b) More flexible
c) More choice
d) More apps
e) Can use on different hardware
f) Can customize
g) Android is the best mobile operating system

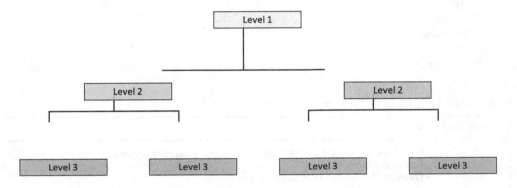

*Figure 9.2* Minto Pyramid Principle outline

6   Check which of the following slide titles in Table XII answer the question 'What's the *So what?*'

*Table XII* Answering the 'So what'?

| Slide title | Answers question |
| --- | --- |
| Number of consumers using mobile payment apps over last ten years. | Yes/No |
| 395% increase in number of consumers using mobile payment apps over last ten years demonstrates market for micro-payment app. | Yes/No |
| The relative market of micro-payment apps | Yes/No |
| Comparison between micro-payment apps in mobile sector | Yes/No |
| Three reasons why the micro-payment app sector is about to explode | Yes/No |
| The advantages of micro-payment apps for mobile wallet users | Yes/No |
| The time taken to establish brand recognition in financial app markets suggests that development lead times need to be at least 18 months in advance of expected gains | Yes/No |
| Loss of market share for banks who failed to establish a mobile app indicates the necessity for app development | Yes/No |

**7a** Match the type of chart to the purpose in Table XIII.

   a) Bar chart
   b) Line chart
   c) Pie chart
   d) Scatter plot

*Table XIII*  Chart types

| Chart type | Purpose of chart |
| --- | --- |
| 1 | To compare categories against a total |
| 2 | To show the relationship (or lack of it) between variables |
| 3 | To compare categories against each other |
| 4 | To show how categories change over time |

**7b.** Match the type of chart to the language associated with it in Table XIII.

   a) Bar chart
   b) Line chart
   c) Pie chart
   d) Scatter plot

*Table XIV*  Language for chart types

| Chart type | Typical language used to title and describe it |
| --- | --- |
| 1 | the relative popularity of . . . , the distribution of market share of. . . , the relative prominence of . . . |
| 2 | the changing landscape of . . . , the growth/decline in . . . , the evolution in . . . |
| 3 | the association between . . . , the correlation between . . . , the link between . . . |
| 4 | the most common . . . , the leading . . . , the principal . . . |

# Works cited

1 J. Bezemer and G. Kress, *Multimodality, learning and communication a social semiotic frame*. Routledge, 2016.
2 D. Biber, "Writing and speaking," in *The Routledge international handbook of research on writing*, 2nd ed. Routledge, 2023, pp. 124–138.
3 M. C. Wolf, M. M. Muijselaar, A. M. Boonstra, and E. H. de Bree, "The relationship between reading and listening comprehension: Shared and modality-specific components," *Reading and Writing*, vol. 32, pp. 1747–1767, 2019. https://doi.org/10.1007/s11145-018-9924-8
4 J. M. Denstadli, T. E. Julsrud, and R. J. Hjorthol, "Videoconferencing as a mode of communication: A comparative study of the use of videoconferencing and face-to-face meetings," *Journal of Business and Technical Communication*, vol. 26, no. 1, pp. 65–91, 2012. https://doi.org/10.1177/1050651911421125
5 M. Markel and S. A. Selber, *Technical communication*, 12th ed. Bedford St. Martins, 2018.
6 B. Minto, *The minto pyramid principle: Logic in writing, thinking and problem solving*. Minto International Inc, 2010.
7 J. M. Lannon and L. J. Gurak, *Technical communication*, 13th ed. Pearson, 2014.

# Chapter 10

# Case studies

## Applied communication in IT

### Note to the teacher/independent learner

The following materials can be used as a self-study opportunity to consolidate the skills developed in the previous nine chapters. They have been designed with the possibility of being used as assessments for a course teaching English for IT, IT, or technical communication skills. The materials employ scenario-based assessments, which aim to situate the IT communication skills within real-world contexts [1, 2]. This approach is intended to help students apply and demonstrate their understanding of the IT communication skills needed and valued in practical, realistic scenarios. Some case studies are shorter and lead to the development of one form of technical communication. Others are more extended, following the SDLC and involve producing several different texts and types of communication.

Teamwork, collaboration, and communication are some of the most prized skills in software development companies [3]. A 2017 study found that the success of a software development team is strongly associated with the quality of the team itself [4]. This chapter provides a range of case studies that are designed to apply the skills and knowledge of the textbook to a series of scenario-based learning tasks. Each task could be used as an assessment in a formalized learning environment, or they could be used to provide more informal training. They are also possible independent tasks for novice IT professionals to develop their communication skills in a variety of scenarios.

To promote the development of all three valued skill sets, it is strongly encouraged to facilitate these tasks as group tasks, according to the guidelines in each section. They are designed to provide students with experience of working in teams to make professional decisions and practice their professional IT communication skills [5, 6].

Read the scenarios carefully. They will provide the necessary context for each writing choice you make in the subsequent tasks.

### Case study 1: conduct a stakeholder profile

#### Scenario:

You work at a medium-sized application design company in your area. Your company is primarily concerned with creating API applications for small businesses to automate and streamline their processes. Your company is currently considering developing a mobile app to facilitate dry-cleaning services. The app will connect potential customers with local dry-cleaning companies, offering scheduling, pick-up and delivery, and payment options.

DOI: 10.4324/9781032647524-10

In order to proceed with the associated tasks, your team will need to brainstorm the following considerations:

- Company name and overview
- Organization structure (see Chapter 3 for ideas) and workplace culture.
- Teamwork and collaboration: if you are completing this in a group, you will need to assign hypothetical company roles to fulfil throughout the duration of this project.

## 1  Audience analysis

Read the scenario at the beginning of this case study. The first task is to consider the audiences you need to reach. Conduct an extensive audience and purpose analysis, using the materials and concepts covered in chapter one.

Your analysis should clearly identify and rationalize the following:

- Different audiences suggested by the scenario – end user, external/internal, etc.
- Primary audience (end user)

  - IT background
  - Connection to the product
  - Needs from the product
  - Values and desires from the product

- Type of audience
- Contexts of use

Next, brainstorm the types of documents that may be associated with the product. Use Table I to list them.

*Table I* Types of audience, associated documents, and contexts of use

| Type of audience | Type of document | Context of use |
|---|---|---|
| e.g. End user | Quick-start guide | Used to help get the product set up and basic usage |

The analysis should be written using a mix of explanatory paragraphs and tables and lists that outline and preview material accordingly. Make sure to group these effectively and use headings to make sections distinct.

## 2  Audience variety

a) What are the different readers/consumers of content suggested by the scenario?
b) Map out the audience, type of information, and content of information. Consider the most effective way to visually convey this information – often a table is an effective way to achieve this.

- Internal/External
- IT knowledge
- Reason for reading/consuming content

*i  Extended writing task*

Write an internal memo to cover the audience analysis that you conducted earlier and to introduce your analysis of the feasibility of this application. Your memo should follow the guidelines as outlined in Chapter 3, Section 3.3. It should be written appropriately according to audience and purpose. Identify which of the three main purposes your memo demonstrates. Follow the layout expectations.

Your memo should follow this structure:

1  Details of record
2  Greeting
3  Purpose statement
4  Context and rationale
5  Closing
6  Identifying information

## 3  Purpose analysis

The second task is to consider what is needed from a technical standpoint. To do this, develop the earlier contexts of use section from the audience analysis in question one in more detail. Your team will need to brainstorm the technical requirements from the perspective of end users.

a) Map out user needs, considering the following stakeholders:

- Businesses who subscribe to your service
- End users who download and use your app
- System administrators who manage the back-end of the app and service

b) Now, write a user story (see Chapter 5, Section 5.2) for each previous stakeholder.

Remember to use the appropriate structure:

**As a** [user type]**, I want to** [perform a function] **so that** [benefit].

Remember to refer to the INVEST approach to ensure your user story fits the expected attributes.

## 4  Task analysis

Once you have a clear picture of your user needs, it is time to connect their needs to the software. This is achieved using several different strategies. Chapter 5, Section 5.2 includes an overview of these and explains the INVEST approach.

- Use Case Diagrams and Descriptions
- Software Requirements Specification (SRS) Document
- API Documentation

*i  Extended writing task*

Since the application your team is tasked with developing is an API, you will create OpenAPI Specification documentation (OAS).

Review the information in Chapter 5, 5.5 API Documentation.

a) Complete the following sections individually, according to your analysis of the project to date:

- **Overview or about** – include the purpose of the API, its features and capabilities, the intended audience, and any related services or APIs. Include a diagram illustrating high-level architecture. Decide on fair usage, include definitions for key terminology, and versioning and feedback information.
- **Authentication and authorization** – Explain sign-up and registration, key/token generation, and safe usage.
- **Status and error codes** – include the meaning of status and error codes.
- **Quick reference guides** – explain how to perform basic tasks and provide a clear understanding of the functionality of the API.
- **References** – explain each endpoint of the API in detail, such as object/resource descriptions, parameters, or request and response examples.

b) Ensure that you include all relevant information and pay attention to formatting that follows the expectations for each section, for example, if a table is expected, or if a diagram requires an explanatory paragraph. Above all, think about your *audience* and their *needs and wants* from this information.

*ii  Extended writing task*

Write an email to your marketing manager to schedule a meeting to contact local businesses to gauge potential interest in the application. Consider your audience needs, what content is likely to be most persuasive, and their level of potential welcome of the message. Follow the guidelines for writing an email in Chapter 3, Section 3.2.

*iii  Optional extension task*

Write an email response as the marketing manager to the email of a classmate.

*iv  Optional extended writing task*

Write a proposal report to outline the results of your audience and task analysis. Refer to Chapter 7, 7.2 for a review of the basics of writing a proposal report. Remember that the proposal is a persuasive document. It should be pitched at the relevant person within your company who can act on your proposal.

Your proposal should include the following:

1  Executive Summary
2  Problem
3  Solution
4  Criteria
5  Strategy
6  Budget

7 Timeline
8 Team Profile
9 Conclusion
10 References

You may make use of the materials already generated in previous tasks to compile this report. Refer to Chapter 7, 7.7 for more information about writing the various sections. Be sure to use IEEE formatting (see Chapter 8) for your in-text citations and end references.

v  *Optional extension task*

Prepare an oral presentation of the components of your proposal report for the senior management of your company. Include the key information, but structure it appropriately for a presentation. Refer to Chapter 9, 9.3 Presentations to clearly structure your presentation. Ensure you have anticipated the type of audience you are likely presenting to, and then adjust your strategy accordingly. Ensure you have correctly identified the main purpose of your presentation. Use a clear structure, choosing one of the strategies outlined in 9.3.

a) Include a slide deck that provides visual support for your presentation.
b) Structure your presentation effectively, using one of the strategies given in Chapter 9, Section 9.3.
c) Choose a delivery method and prepare accordingly. Ensure each group member has an equal contribution of the presentation.
d) Use the presentation checklist to ensure that you have included the key components.

### Suggested assessment tasks

- Audience analysis and internal memo
- Task analysis – write the user stories associated with the task analysis. Perform a group self-assessment using the criteria of INVEST to evaluate the strength of the user stories (see Chapter 5, 5.2).
- API Documentation of relevant sections
- Email to internal marketing manager
- Proposal report
- Oral presentation of proposal to board
- Team presentations – student groups present their audience and purpose analysis to rest of class, presenting findings and explaining the rationale informing their findings.
- Peer review of group presentations using presentation checklist.
- Peers provide feedback to one another and write reflective comparisons on the similarities and differences of their own analysis and that of their peers.

## Case study 2: redesigning a company website

### Scenario:

You are a technical writer working for a medium-sized web application design company. Your company is primarily concerned with creating API applications for small businesses to automate and streamline their processes. Recently, a potential client has approached you to update and improve their website design. The client is a small business owner who runs a local bakery. She loves dogs and has named the bakery after her cocker spaniel, Lola. The client wants the website to be visually appealing, easy to navigate, and optimized for mobile devices.

## 1   Website design

Design a website homepage for the client's bakery that adheres to the ten most important principles of design: unity, balance, alignment, hierarchy, emphasis, proportion, white space, repetition, movement, and contrast. Your design should be visually appealing and easy to navigate. It should reflect the nature of the business and the business owner's personality. You should also ensure that the website is optimized for mobile devices.

## 2   Modifying website design

Which design features would you change if you were designing the web content for

- A bakery in Jeddah
- A bakery in Shanghai
- A bakery in Paris
- A bakery in Mumbai
- A pet-friendly café in Tulum

Suggested Assessment Tasks

- Wireframe of the website design that includes

  - A brief explanation of how the design meets the requirements of the ten principles of design
  - A list of any images or other media that would be used on the site
  - A written reflection that justifies your design's suitability for your client's values and needs

- Oral presentation of wireframe to class

## Case study 3: reviewing automated messages
### Scenario:

Your company wants to use an external project management tool to monitor and manage IT workflow processes and ticketing. While this will automate workflow processes, it is important to ensure the messages that are automatically sent align with the local culture and that the associated knowledge base is up-to-date.

## 1   *Automated message analysis*

a) Conduct a survey of automated messages needed.
b) Write the body of the automated messages. Include details relevant to your context and ensure the messaging fits the expectations of the context where you work.
c) Conduct a review of the knowledge base. What are the gaps? What information is needed to fill them?
d) Is there an IT policy that needs to be modified on the basis of this audit? Do modification records need to be updated?

### i   *Optional extended writing task*

Write a brief status/progress report to your supervisor, providing an overview of your results. This report should be relatively informal, and it should identify the following:

a) What is completed (which automated messages have been reviewed/written)
b) What still needs doing (which automated messages need to be written or updated, which gaps in the knowledge base have been identified)
c) Are there any challenges that you have encountered?
d) What is the overall view of your audit?

Make sure the report follows the guidelines for effective progress reports in Chapter 7, Section 7.5.

### ii   *Optional extended writing task*

Write a standard operating procedure (SOP) outlining how your company responds to basic IT helpdesk requests, such as forgotten passwords. Ensure you include the sections outlined in Chapter 6, Section 6.2 about policy documents and standard operating procedures.

### iii   *Optional extended writing task*

Your team has been asked to create a self-access tutorial for employees to use to reset their password when necessary.

1   Use the procedure from standard operating procedure to build a script to narrate your screen recording

   a   Consider audience needs to choose your words carefully

2   Choose a screen recording technology to create the screencast
3   Practice completing the task while narrating the action
4   Record your screencast
5   Test your screencast by having a different team member follow your recorded tutorial

### *Suggested assessment tasks*

- Sample automated messaging text
- Progress report outlining results of audit

- Standard Operating Procedure (SOP) for dealing with forgotten passwords
- Self-access tutorial video

## Case study 4a: developing a micro-payments app

### Scenario:

You work at a medium-sized application design company in your area. Your company is primarily concerned with creating API applications for small businesses to automate and streamline their processes. A current project that your company is considering is to develop an app to facilitate small payments such as tipping. The app will enable customers to tip service providers without requiring banking details.

Several local businesses have expressed interest in the app. They have agreed to sign on for a six-month period, provided the app can be up and running within six months. Your lead developer has asked you to start the project. Before you can begin, you need to decide on a project management model. So you can make the best choice of a model, you also need to consider how long the project is likely to take, since the length of time is an important factor to determining which project management model is best.

Refer to the comparative analysis in Chapter 4, 4.1.

The chapter compares waterfall and agile development models. A third option that is not compared in Chapter 4 is the spiral model.

## 1   Project management model comparative analysis

a) Conduct some research to learn about how spiral development compares to waterfall and agile. Use the categories found in Chapter 4, Section 4.1.
b) Using information from the scenario, from your own research and applying safety and legislative considerations of your local context, evaluate this project across the following three options. Use Table II to facilitate this comparison.

*Table II* Project management model comparison

|  | Waterfall | Agile | Spiral |
|---|---|---|---|
| **Benefits** | | | |
| **Drawbacks** | | | |

c) Choose a project management model to use for this project. Write a rationale for your selection using the information in Table II.

*i   Optional writing task*

Compile the research and comparison into a short report. Include a clear written rationale that explains how your selection is best. Refer to the information used in Table II to conduct the comparative analysis.

## 2   Time, scheduling, and cost management documentation

a) Time and Scheduling Management Documentation
   Once you have selected a model, complete a Gantt chart (see Chapter 4, 4.3) outlining the key processes of the software development life cycle. Estimate the amount of time your team will need for each stage. Be clear about the delegation of tasks. Identify any areas of potential overlap, as they could be problematic later. You should also create a strategic roadmap for the process. Remember to adhere to the writing guidelines outlined in Chapter 4.

b) Cost Management Documentation (See Chapter 7, Section 7.7 Money management)
   The next stage involves estimating how much this project will cost for your company. Decide whether to use analogous estimation or parametric estimation to complete a table documenting your anticipated expenses. See Chapter 7, 7.7 for more information and an example. Don't forget to reference each figure and include explanations written for a non-specialist audience.

c) Product Requirement Documentation
   Next, your team need to consider what software requirements are needed. Follow the guidelines in Chapter 5, 5.2 to create a product requirement document (PRD). Look at the examples and decide which of the sections are necessary for this context. Follow the guidelines for layout and make sure your PRD clearly defines the problem, outlines the proposed solution, and lists the objectives of the project. Remember to clearly provide the scope of your project and include a section for questions.

## 3   Feasibility study

Now you have enough information to start to put together a feasibility study about the value of the project. You may use much of the information you have already collected to help create the feasibility study, but make sure you repurpose it according to the expectations of a feasibility study. Refer to the guidelines in Chapter 7, Section 7.4. Make sure you identify the relevant stakeholders in your context.

Ensure your study provides a clear answer to each of the questions in the purpose section:

1   Is it possible?
2   Is it worth it?

Your feasibility study should include the following sections:

- Executive summary
  - What's in the report?
- Project description
  - What project is being assessed for its feasibility?

- Obstacles

  - What obstacles need to be overcome to make the project happen?

- Benefits and drawbacks

  - What are the benefits of this project to our organization?
  - What are the drawbacks of this project to our organization?

- Recommendation

  - Is the project feasible?

- End matter

  - What evidence did you use to make your recommendation?

Refer to guidelines in Chapter 7, Section 7.7 Common report features, and Section 7.8 Language focus for further advice and ideas.

### Suggested assessment tasks

- Comparative analysis
- Written rationale
- Completed time, scheduling, and cost management documentation
- Product requirement document
- Feasibility study

## Case Study 4b: developing a micro-payments app

### Scenario:

Now that your team have agreed on a project management model and you have scheduling and cost management plans, begin the process of designing the application.

## 1  Use case diagram

The first stage is to create a visual representation of how the application will work. To do this, you will create a use case diagram (UCD). Remember: this diagram illustrates how your planned application will work

### i  Optional writing task

Create a UCD of how the application will work. Use the space to diagram how the application will facilitate the interaction between a customer who wishes to use the application to tip a service provider.

Consider

- The actors
- The use cases
- System boundary
- Relationships

Customer                                                    Service Provider

*Figure 10.1* Use case diagram for micro-payment application

Now, create a UCD in Figure 10.1 that illustrates how using the application connects the end user (customer) with a potential service provider.

## 2    Software requirements specification document

The next stage is to create a more technical document for the development of the application. Using the product requirement document and the use case diagram, develop a software requirement specification (SRS) document.

To do this, you will need to conduct some research into how apps that deal with financial transactions are usually developed.

### i    Extended writing task

Create your Software Requirements Specification (SRS) document. Refer to the advice in Chapter 5, 5.3 and ensure your SRS includes each of the following sections:

1 **Introduction** – provide the purpose, the intended audience, the scope of the project, key definitions, acronyms, and abbreviations. Include links to the references used to generate the report.
2 **Product description** – include the product perspective and functions, user classes, and characteristics. Outline the operating environment and constraints. Include any design and implementation constraints. List the user documents that need to be included with the software. Any assumptions should be listed.
3 **System features and requirements** – describe the features and the functional requirements for the software, separate out the non-functional parameters as well.
4 **Use case diagram** – create a graphical representation of how the different components of the intended application will interact. Include description of the entities, their relationships, and attributes. If you completed the previous optional writing task, you can include it here.

Ensure your SRS is clearly formatted, well-researched, and supported by references. Be sure to use IEEE formatting and refer to Chapter 8, Sections 8.3 and 8.4 for guidance on in-text citations and end-reference formatting.

### ii    Extended writing task

Your team has been contacted by the specialist designer assigned to develop the user experience design (UX). S/he has some questions arising from the SRS document created by your team. To

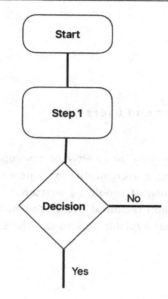

*Figure 10.2* User flow diagram

clarify the project from the UX perspective, create a user flow and a wireframe so that s/he is better able to visualize your idea and begin work. A user flow diagram is similar to a use case diagram, but it creates a step-by-step decision tree diagram of the process. Use Figure 10.2 to brainstorm the steps taken by the end user to facilitate the micro-payment.

Refer to Chapter 5, Section 5.4 UX Design Documentation for more details.

### iii   Optional extended writing task

Now your team needs to consider the project from the perspective of the end user. Choose one of the user documents listed in your SRS document. Prepare the document using the guidelines found in Chapter 6.

### iv   Optional extended writing task

Create a tutorial for the basic use of sending a micro-payment to a service provider.

1   Use the user flow diagram to break the process down.
2   Create a script for narrating the tutorial.
3   Use a relevant screen recording application to record it.

### Suggested assessment tasks

- SRS document
- Use case diagram and description

    - It is optional to separate this from the SRS and create a standalone assignment.

- User flow
- Wireframe
- Tutorial of basic use

## Case study 5: writing for end users

### Scenario:

You are a technical writer working for a software development company. Your team has been tasked with creating a user manual for a new software product that the company is launching. The manual should be written in plain language and should be easy to understand for non-technical users. You have been assigned to write the section of the manual that explains how to use the software's search function.

## 1    Extended writing task

Write a user manual section that explains how to use the search function of the software in plain language. Your section should be no more than 500 words long and should include screenshots and diagrams where necessary. Consider where users may find additional help and resources and provide this information appropriately.

Refer the guidelines for writing about procedures in Chapter 6, Section 6.2. Remember also to account for contexts of use. Format the guide appropriately, including action-oriented headings, numbered lists, and careful application of design principles like white space and consistency to help your reader follow the procedure. Refer to Chapter 2 for more information about applying the design principles.

## 2    Optional extended writing task

Your team has been asked to create a multimodal video tutorial that in-house employees can use.

1  Use the procedure from the extended writing task to build a script to narrate your screen recording.

   a  Consider how to verbally preview a task.
   b  Signpost how smaller actions combine to accomplish larger tasks.

2  Choose a screen recording technology to create the screencast.
3  Practice navigating across the platform while narrating your actions.
4  Record your screencast.
5  Test your screencast by having a different team member follow your precise directions.

## Further case studies

*Get more case studies, examples, materials and quizzes for the whole book at* https://english-foritcommunication.com.

# Works cited

1  T. Beaubouef, R. Lucas, and J. Howatt, "Reviewed papers: The UNLOCK system: Enhancing problem solving skills in CS-1 students," *ACM SIGCSE Bulletin*, vol. 33, no. 2, 2001.

2  R. Iqbal and P. Every, "Scenario based method for teaching, learning and assessment," in *Proceedings of the 6th Conference on Information Technology Education*, Newark, NJ, USA, Oct. 2005, pp. 261–266.

3  Y. Lindsjørn, D. I. K. Sjøberg, T. Dingsøyr, G. R. Bergersen, and T. Dybå, "Teamwork quality and process success in software development: A survey of agile development teams," *Journal of Systems and Software*, vol. 122, pp. 274–286, Dec. 2016. https://doi.org/10.1016/j.jss.2016.09.028

4  E. Weimar, A. Nugroho, J. Visser, A. Plaat, M. Goudbeek, and A. P. Schouten, "The influence of teamwork quality on software team performance," arXiv preprint arXiv: 1701.06146 2017.

5  H. Koppelman and B. van Dijk, Creating a realistic context for team projects in JCI, in *Proceedings of the 11th Annual SIGCSE Conference on Innovation and Technology in Computer Science Education*, Bologna, Italy, 2006, pp. 58–62.

6  H. Yuan and P. Cao, "Collaborative assessments in computer science education: A survey," *Tsinghua Science and Technology*, vol. 24, no. 4, pp. 435–445, Aug. 2019. https://doi.org/10.26599/TST.2018.9010108

# Index

abbreviations 164–165
acceptance criteria 89–91
access control documentation 125–129
agile development 70
alignment (design principle) 24–27
API documentation 95–99
appendices 168
artifacts, guidelines for creating, xv–xvi
audience: for API documentation 96; analysis of 5–7, 191–192; for end-user documentation 109–110; for presentations 174–176; for proposals 136; for quality assurance documentation 99–100; for recommendation reports 138; for software requirements specification 92; types of in IT 7

balance (design principle) 23–24
bar charts 181

charts and graphs 181–182
citations: how to cite 156–157; placement of 156–157; styles of 157–158; when to cite 158; when not to cite 158–159
coding standards documents 79–80
communication: intercultural considerations 49–51; internal vs external 47–49; multimodal 172–174
contrast (design principle) 36–37

data visualization 40–44
design principles 22–44
documentation: for end-users 108–131; process vs product 70–71; for quality assurance 99–104; for system administrators 125–131; types of SDLC 69–71; value of SDLC 69–70

email 51–56
emphasis (design principle) 29–31
end-user documentation 108–124

entity relationship diagrams 41–42
evaluation reports 141–142
executive summaries 147–148

feasibility reports 138–140
figures 167–168

Gantt charts 72–74
grammatical clarity 103

headings 38, 168
hierarchy (design principle) 27–29

IEEE referencing 154–169
instruction manuals 111–116
in-text citations 156–159

kanban boards 74–75

language: for presentations 185; for report writing 148–150; for user documents 124–125

memos 56–61
multimodal communication 172–174

OpenAPI Specification (OAS) 96

presentations 174–185
progress reports 140–141
proportion (design principle) 31
proposals 136–137

quality assurance documentation 99–104
quick-start guides 111, 116

recommendation reports 137–138
referencing, IEEE 154–169
release roadmaps 78–79
repetition (design principle) 32–35

reports: common features of 142–148; types of 134–135
roadmaps 75–79

software requirements specification (SRS) document 92–94
standard operating procedures (SOPs) 120–122
strategic roadmaps 76–78
system admin documentation 125–131

tables 166–167
technology roadmaps 77–78
tenses 103–104

unity (design principle) 22–23
use case diagrams 42
user experience (UX) design documentation 94–95
user flow diagrams 42–44
user stories 88–89

video conferencing 174–175

white space (design principle) 31–33
work breakdown structures 71–72
working papers 80

Printed in the United States
by Baker & Taylor Publisher Services